Encyclopedia of Human Body Systems

Encyclopedia of Human Body Systems

VOLUME I

Julie McDowell, Editor

GREENWOOD

AN IMPRINT OF ABC-CLIO, LLC
Santa Barbara, California • Denver, Colorado • Oxford, England

Copyright 2010 by ABC-CLIO, LLC

Library of Congress Cataloging-in-Publication Data

McDowell, Julie.
  Encyclopedia of human body systems / Julie McDowell.
    p. cm.
  Includes bibliographical references and index.
  ISBN 978–0–313–39175–0 (hard copy : alk. paper) — ISBN 978–0–313–39176–7 (ebook)
1. Human physiology—Encyclopedias. I. Title.
QP11.M33   2011
612.003—dc22   2010021682

ISBN: 978–0–313–39175–0
EISBN: 978–0–313–39176–7

14 13 12 11 10   1 2 3 4 5

This book is also available on the World Wide Web as an eBook.
Visit www.abc-clio.com for details.

Greenwood
An Imprint of ABC-CLIO, LLC

ABC-CLIO, LLC
130 Cremona Drive, P.O. Box 1911
Santa Barbara, California 93116-1911

This book is printed on acid-free paper ∞

Manufactured in the United States of America

Contents

About the Editor
and Contributors

About the Editor

Julie McDowell is a science and health care journalist based in Washington, D.C. She is the author or coauthor of several books in the sciences, including *The Nervous System and Sense Organs*, *The Lymphatic System*, and *Metals*. She holds an MA in nonfiction writing from Johns Hopkins University in Baltimore, Maryland.

About the Contributors

Amy Adams has written science news and features for publications including *The Scientist*, *Discovery Channel Online*, *CBS HealthWatch*, and *Science*, among others. She holds an MS in developmental genetics from Cornell University and is a graduate of the Science Communication Program at the University of California, Santa Cruz.

Kathryn H. Hollen is a science writer who has worked extensively with the National Institutes of Health, especially the National Cancer Institute, as well as many other organizations devoted to the life sciences.

Evelyn Kelly is an independent scholar and adjunct professor at Saint Leo University, Florida Southern College, and Nova University. She has written more than 10 books and over 400 journal articles. She is a member of the National Association of Science Writers and past president of American Medical Writer's Association, Florida chapter.

Leslie Mertz is a biologist at Wayne State University. She is the author of *Recent Advances and Issues in Biology* (2000), and former editor of the research magazine *New Science*.

David Petechuk is an independent scholar. He has written on numerous topics including genetics and cloning and has worked with scientists and faculty in areas such as transplantation and psychiatry.

Kelli Miller Stacy is a board-certified life sciences editor and the founder of NEWScience Inc. where she is a feature writer on health and medicine for clients such as *The Scientist* and *Popular Science*. She is the winner of the American Medical Writers Association's Eric Martin Award for excellence in medical reporting.

Stephanie Watson is a professional medical writer and editor. She is the author of *Medical Tourism: A Reference Guide* (2011).

Michael Windelspecht is Associate Professor of Biology at Appalachian State University. He is the author of two books in Greenwood's *Groundbreaking Scientific Experiments, Inventions, and Discoveries through the Ages* series and editor of Greenwood's *Human Body Systems* series.

Introduction

The *Encyclopedia of the Human Body Systems* provides an overview of the physiology of the major organ systems of the body. For the purposes of this book, a system is defined as an organ group that works to perform a function for the body.

This book looks at 11 systems, with each chapter dedicated to exploring a specific system. The first chapter of the book provides an overview of the human body, including a look at its cellular foundation, chemical composition, and how the body is organized as well as an explanation of anatomical terms used to describe different areas of the human body. The subsequent chapters then focus on each of the systems. The purpose of this organization is so that readers can understand the basics of the human body and then go on to learn about each system.

The primary organs and functions of each system that are covered in this book's chapters are outlined below, as well in Table A. It's important to note that while each system does have its own distinct function, the systems interact with each other and rely on the organs of other systems to function properly.

- **The Circulatory System:** The major organ is the heart, which functions with the help of blood vessels. The circulatory system's major responsibility is to ensure that the blood transports vital substances like oxygen throughout the body to the heart and other systems' organs, including the liver, stomach, and brain.

- **The Digestive System:** The major organs of this system are the mouth, esophagus, stomach, and small and large intestine, while the secondary organs are the liver, pancreas, salivary glands, and the gall bladder. The digestive system's primary role is to turn substances coming into the body—such as through food and water—into fuel for its cells. This process is called metabolism, and is part of the larger, more comprehensive digestive process.

- **The Endocrine System:** The primary organs of this system (more commonly called glands in the case of the endocrine system) are the hypothalamus, pituitary, thyroid, parathyroids, adrenal, pancreas, and sex glands. The endocrine system is one of the body's two communication hubs, the other being the nervous system. In this system, the communication is carried out through hormones, which are chemicals that travel through the bloodstream that prompt stimulation and inhibition of nerve impulses.

- **The Integumentary System:** Major organs are the skin and accessory organs, including hair and nail follicles. The primary job of this system is to protect the body, as well as detect changes that affect the body using its sensory receptors.

- **The Lymphatic System:** The primary organs of this system are the bone marrow and thymus, while the secondary organs are the spleen, tonsils, adenoid, Peyer's patches, and appendix. The lymphatic system's job is to protect the body against toxins and other potentially harmful substances that can cause illness and disease. While not an organ, one of this system's most important components are lymphocytes. These specialized cells detect organisms that might be harmful to the body and then prompt an immune response to drive them out of the body.

- **The Muscular System:** Three types of muscles make up this system: the skeletal muscle, which helps the body to move; the smooth muscle, which is associated with the internal muscles; and cardiac, which works to help the heart to function. The movement of each of these muscles is determined by direction they receive from different areas of the body. Specifically, the autonomic nervous system controls the smooth and cardiac muscles and the central nervous system controls the central nervous system.

- **The Nervous System:** The primary organs of the nervous system are the brain and spinal cord. Through the nervous system's two divisions, the central nervous system (CNS) and the peripheral nervous system (PNS), stimuli and other information is processed into the form of reaction and activity. In addition to the endocrine system, the nervous system is known as one of the body's key communication centers. This communication is done through nerve impulses that travel through the body's nerve fiber system.

- **The Reproductive System:** The role of this system is to carry out the process of sexual reproduction, which is necessary to continue the human species that contains genetic information that determines a person's physical characteristics, but also their resistance to certain mutations also ensures the future of the species. This system is unique because there are distinct sets of organs for females and males. In addition, there are internal and external organs in the male and female, with the internal organs located inside the body and external organs located outside the body. Some examples of the internal female reproductive system include the cervix, vagina, and fallopian tubes; external organs include the mammary glands or breasts. For males, internal organs include the ejaculatory ducts and urethra; external organs include the penis and testicles.

- **The Respiratory System:** The primary organs are located in two areas: the upper and lower respiratory tracts. The upper tract contains the nose and nasal cavity (also known as the nasal passage), the pharynx (or throat), and the larynx (or voice box). The lower respiratory tract contains the trachea (windpipe), the bronchi, the alveoli, and the lungs. This system's major responsibility is to control and regulate the breathing process, which involves moving air into and out of the lungs.

- **The Skeletal System:** The main components of the skeletal system are the body's 206 bones, as well as the tendons, ligaments, and joints that connect the bones to the muscles, controlling their movement. The skeletal system is similar to the integumentary system in that it plays a protective role. Like the skin, skeletal bones

TABLE A
Human Body Systems

System	Some examples of organs
Circulatory	Heart
Digestive	Stomach, small intestine
Endocrine	Thyroid
Lymphatic	Spleen
Muscular	Cardiac muscle, skeletal muscle
Nervous	Brain, spinal cord
Reproductive	Testes, ovaries
Respiratory	Lungs
Skeletal	Bones, ligaments
Urinary	Bladder, kidneys
Integumentary	Skin, hair

Note: This organ list just provides some examples and is not meant to be complete.

protect tissues and organs, as well as the other important internal organs from the body's other systems.

- **The Urinary System:** This system's primary organs are the kidneys, ureters, bladder, and urethra. The urinary system's primary job is to rid the body of waste, through producing urine and other elimination processes. In addition, this system ensures that the body's fluid system is balanced in terms of acidity.

Each chapter in this volume opens with a listing of interesting facts pertaining to each system, as well as the topics or highlights that will be covered in each chapter. Do not feel the need to memorize these highlights; it is just a way of alerting the reader to the focus of the chapter. Each chapter also contains a "Words to Watch For" section. This listing is an overview of the important terms that the reader can expect to learn about in each chapter. Many of these terms are also highlighted in **bold** in the chapter and then defined in the glossary, which is located at the back of both volumes for the reader's convenience. Throughout the book, tables and appropriate graphics will be used to help illustrate some concepts in a visual manner.

This book will be useful for students with various backgrounds in understanding human biology. Presented in a clear and concise manner, this book explains basic concepts of the human body—such as cell and blood composition—and then builds on this foundation to explore more complex concepts, such as nerve impulse transmission and respiratory processes. Whether readers open this book with no grasp of or a firm comprehension about the workings of the human body, they will broaden and deepen their understanding about a subject that has confounded some of the greatest minds in history, including Hippocrates, Aristotle, and Michelangelo. The mysteries of the human body continue to perplex scientists, medical experts, and the general public. However, these minds—from the past and present—found learning about the human body exhilarating and vital in order to improve the quality of life and find cures and treatments for diseases and disorders. But the one thing necessary to explore these mysteries is learning and studying not only the complexities of each of the human body systems, but also the basics—what are they composed of, and how do they function? The *Encyclopedia of the Human Body Systems* is a key resource for this information, and therefore, an important tool in gaining a comprehensive understanding of the human body.

Julie McDowell
Washington, D.C.

1

The Building Blocks of the Human Body

Julie McDowell

Interesting Facts

- It's estimated that over a lifetime, humans produce a total of 10,000 gallons of saliva. While saliva is necessary for us to taste our food, we cannot taste it before some food dissolves in it.

- There are approximately 100 trillion cells in the adult human body.

- There are approximately 31 billion base pairs in the human genome.

- In a single cell, there are six billion "steps" of the DNA or body's genetic code. If stretched out, it would measure six feet, but it is twisted up in a coil in the nucleus of the cell, where its diameter measures only 1/2500 of an inch.

- A mammal's lifespan is typically determined by its size. But because humans have developed ways to protect themselves, their lifespan is longer. More specifically, a man's lifespan should be 10 to 30 years—somewhere between a goat and a horse—but it's actually an average of 74.7 years in the United States.

- Thanks to the body's complex and extensive circulatory network, between 40 and 50 percent of body heat is lost through the head. This is why hats keep the body warm in winter—they keep the heat in the body.

- Compared to other cells in the body, brain cells live the longest, with some living an entire lifetime.

- Every second, the human body destroys—as well as produces—15 blood cells.

- Approximately 90 percent of the body is made up of four elements: oxygen, carbon, nitrogen, and hydrogen.

- The brain is especially reliant on oxygen to function. In fact, it uses more than 25 percent of the body's oxygen supply.

Chapter Highlights

- Levels of the human body

- Chemical elements of the body

- Cells and tissues

- Organs and organ groups

- Body cavities

- How is the body organized?

- Key anatomy terms

Words to Watch For

Adiopose tissues	Homeostasis	Pelvic cavity
Cells	Hormones	Peritoneum
Cell membrane	Inorganic chemicals	Plane
Cell organelles	Median sagittal plane	Serous membranes
Cell respiration	Meninges	Solvent
Cranial cavity	Mesentery	Spinal cavity
Cytoplasm	Nucleus	Tissues
Dorsal cavity	Organic chemicals	Transverse plane
Enzymes	Organs	Ventral cavity
Frontal plane	Organ systems	

Introduction

The human body is a composed of a complicated organizational structure that becomes increasingly complex. As the complexity increases through the various levels of the human body, the behavior of each level is by the function and operation of the previous—or simpler—level of the body. But just because the human body is complicated and complex does not mean that many aspects cannot be studied and understood. While many mysteries still surround the human body, there is much that we know and understand. The key to this understanding is starting at the most basic level, and that is the chemicals of the human body. This chapter will begin with a brief overview of the levels—from the chemical to the organ system level—and then go into more details about certain aspects of each of the levels, such as cells, tissues, and membranes (see Table 1.1). Of course, more details on all of these levels are incorporated throughout *The Encyclopedia of Human Body Systems*.

Chemical Components of the Body

The human body is often described as a large vessel containing chemicals that are constantly reacting with one another. There are two kinds of chemicals that make up the human body: **inorganic** and **organic chemicals**. Although there are some exceptions, inorganic chemicals are primarily molecules that are made up of one or two elements that are not carbon. Water (H_2O) and oxygen (O_2) are examples that are important for the human body to function, as are iron (Fe), calcium (Ca), and sodium (Na). One exception is carbon dioxide (CO_2)—even though this contains

TABLE 1.1
The Structural Organization of the Human Body

Level 1	Chemical
Level 2	Cellular
Level 3	Tissue
Level 4	Organ
Level 5	Organ system
Level 6	Organism

SIDEBAR 1.1
Atoms: Chemical Foundation of the Body

The simplest chemical, in living as well as nonliving matter, is the element, and the simplest element is the atom. In fact, the atom is the smallest unit in the body's chemical system, which is its simplest level. Therefore, the atom really is the most basic element of the human body. The four most common elements in the human body are carbon (C), oxygen (O), hydrogen (H), and nitrogen (N).

Each atom is made up of a center or nucleus. The nucleus also contains protons and neutrons. Floating around the nucleus at rapid speeds are electrons. While each proton is positively charged, the neutrons are neutral. The electrons, however, are negatively charged. In a regular atom, the number of protons and neutrons are equivalent, therefore the electrical charge is neutral. One example is the carbon atom, which has six protons and six neutrons in its nucleus, while six electrons are in orbit around the nucleus on paths known as energy levels.

Electrons are also the way that atoms bond to one another to form molecules. A molecule is made up of tightly bound atoms that behave as a unit. However, the number of electrons in the body will determine how the atoms will bond. The bonding also depends on the arrangement of the electrons.

---

carbon, it is inorganic. Information about the body's chemical components is also covered in detail throughout this book, but below is a brief introduction. (See Sidebar 1.1 for a brief explanation of the importance of atoms in the chemistry of the body.)

The other category of chemicals important to the body is organic chemicals, which contain the two elements hydrogen and carbon. These chemicals include fats, proteins, carbohydrates, and nucleic acids.

Key Inorganic Compounds

As noted earlier, water, oxygen, and carbon dioxide are among the body's most important chemical compounds.

Water

Water, which scientists estimate comprises approximately 60 to 75 percent of the human body, is important for three primary reasons: it acts as a **solvent**, it acts as a lubricant, and because it changes temperature slowly, it is vital in regulating the body's temperature. The first reason—because it is a solvent—means that many substances can dissolve in it, which allows nutrients and other vital components to be transported throughout the body. This is because the body's main transportation systems, blood, are largely composed of water. Therefore, substances like glucose (which comes from food) that are needed for energy in the body can be dissolved in the blood and then delivered to the heart and cells. Another important function is the elimination of waste. Materials that the body does not need, called waste products, are dissolved in the water component of urine and then flushed out of the body through the urinary system.

In addition to acting as a solvent, water is a lubricant, which means it prevents friction between the various surfaces inside the body, such as bones and blood vessels. One example of water acting as a lubricant is in the digestive system. One of the fluids present in this system is the fluid **mucus**, which, like blood, is primarily made up of water. Mucus enables food to move through the intestines and be digested, providing the body with fuel.

The third function of water in the body is that of a temperature regulator. The temperature of water does not change quickly—it has to absorb a lot of heat or lose a lot of heat before the temperature increases or drops. This property enables the body to stay at a fairly constant temperature. In addition, water does an important cooling job for the body in the form of perspiration. When the body is absorbing an excess amount of heat, sweat forms on the skin, which allows the heat to escape the body without damaging any cells.

While water is always moving throughout the body, it is categorized into two types, depending on its location. The two types are intracellular fluid (ICF) and extracellular fluid (ECF). ICF is the water located within the cells. This water makes up approximately 65 percent of the body's total water supply. ECF is the remaining 35 percent of the body's water and is the water outside of the cells. There are three types of ECF—plasma, lymph, tissue fluid (also known as interstitial fluid), and specialized fluid. Plasma is water found within the blood vessels; lymph is water found in lymphatic vessels; and tissue fluid is water found in the small spaces that exist between cells. Specialized fluids include water that performs a specific function, depending on the system in which they are located; examples include cerebrospinal fluid around the spinal cord in the nervous system, and aqueous humor in the eye.

Oxygen and Carbon Dioxide

We are continuously inhaling the gas known as oxygen—without it, we would die. Oxygen plays a vital role in breaking down nutrients such as glucose that need to be transported to various locations to provide the body with energy. This process is known as **cell respiration**. The energy produced through cell respiration is contained in a molecule that is called ATP, which stands for adenosine triphosphate. ATP can be thought of as the fuel required for various cellular processes to occur throughout the body.

In addition to producing ATP, cell respiration also produces carbon dioxide. So while oxygen is inhaled, carbon dioxide is exhaled, and is considered a waste product. It is exhaled because it is a waste product of cell respiration. Like other waste products that are covered throughout this book, carbon dioxide must leave the body. If carbon dioxide builds up in the body, it can disrupt the chemical balance in the body. This can cause acidosis, when fluid becomes too acidic, which can result in calcium deposits in the body's soft tissue. Carbon dioxide buildup in the body is toxic to the heart (see Sidebar 1.2).

Key Organic Compounds

As indicated earlier in this chapter, some of the most important organic compounds in the body include carbohydrates, lipids, proteins, and nucleic acids. These compounds are explained in detail in Chapter 3, which covers the digestive system, but below is a brief introduction.

Carbohydrates

These compounds are made up of carbon, hydrogen, and oxygen, and its main function is to be the body's energy source. There are four types of carbohydrates, all of which are forms of saccharide, which means sugar. The four types are monosaccharides, disaccharides, oligosaccharides, and polysaccharides. The prefixes refer to how many sugars are in the compound.

SIDEBAR 1.2

Cell Respiration: How the Body Produces Energy

The body relies on the chemical reaction known as cell respiration to produce energy, and involves both oxygen and carbon dioxide. The chemical reaction is spelled out in the following equation:

$$\text{Glucose } (C_6H_{12}O_6) + 6O_2 \rightarrow 6CO_2 + 6H_2O + ATP + \text{heat}$$

Glucose comes from food, which is broken down with the help of oxygen. This reaction results in carbon dioxide, which, because it is a waste product, is exhaled. Water, energy (ATP), and heat are also produced. In addition to carbon dioxide, these other substances perform key functions in the body. Water becomes part of the body's fluid system, acting like a solvent and lubricant, while the heat helps to maintain the body's temperature. ATP is energy used for processes vital for the body to function, including digestion and muscle contraction.

Monosaccharides are the simplest sugar compounds, because they only contain one sugar. Disaccharides are composed of two monosaccharide compounds, while oligosaccharides are made up of anywhere between 3 and 20 monosaccharides. Finally, polysaccharides are composed of thousands of monosaccharide compounds.

Lipids

Like carbohydrates, lipids are also composed of carbon, hydrogen, and oxygen. Some lipids also contain phosphorus. Three types of lipids include true fats (or just fat), phospholipids, and steroids.

True fats are made up of a molecule called glycerol, and between one and three fatty acid molecules. These fats are where the body stores excess energy produced by food in the form of calories. If a body does not use all of the calories, it is then converted to fat and stored in the body's adipose tissues. These tissues are located between the skin and muscles.

Triglyceride is one type of true fat. It is when one glycerol is bound to three fatty acid molecules. This is the type of fat often found in highly processed foods, and often found in two forms—saturated and unsaturated

SIDEBAR 1.3
Saturated versus Unsaturated Fat

There are two types of true fats—saturated and unsaturated fats, with a diet high in saturated fats, as well as cholesterol, getting a lot of blame for contributing to heart disease. Heart disease, or atherosclerosis, occurs when there are deposits of cholesterol on the walls of the coronary arteries that supply blood to the heart muscle. These deposits cause the arterial walls to narrow, which then inhibits blood flow through the arteries. If the deposits become too great and blood is blocked from the heart, a clot forms and can result in a heart attack.

Saturated fats are typically in a solid form, and are found in food derived from animals such as beef and pork, chicken, eggs, and cheese, as well as some oils, including coconut and palm oils. Unsaturated fats are also found in oils, but those that are plant-based, including corn, sunflower, safflower, and certain fish oils.

When saturated fats are broken down by the liver, they are used to synthesize cholesterol, which cause the blood cholesterol level to increase. But unsaturated fats behave differently; when they are broken down by the liver, they do not increase the body's cholesterol. In fact, there is evidence that some unsaturated fats may actually help lower cholesterol and prevent atherosclerosis in the body by inhibiting cholesterol from being produced by the liver. This is why a diet that is high in unsaturated fats and low in saturated fats is believed by scientists and medical experts to help prevent heart disease. However, there are other contributing factors to atherosclerosis, including heredity and risk factors such as smoking, as well as being overweight. Therefore, eliminating these contributing factors, through quitting smoking, and exercising to lose weight, help to reduce the risk of heart disease. It certainly helps to eat a healthy diet—to help us maintain a reasonable weight.

fats—on nutrition labels. (See Sidebar 1.3 for more information about saturated and unsaturated fats).

Another type of lipids is phospholipids, which are diglycerides that are bonded to a phosphate molecule. Unlike fats, phospholipids do not store energy. They are a part of a cell's structure; they form part of the cell membrane known as lecithin. They also are an integral part of the body's nervous system, in that they help to form the cell's myelin sheath that protects neurons.

Like phospholipids, steroids form part of the cell membranes. But they also have other functions. One of their primary functions is that they are processed by the liver into bile salts, which are used during the digestion to process fats. In addition, steroids are also involved in **hormones** related to male and female reproductive organs—testes (men) and ovaries (women).

Proteins

Comprised of building blocks known as amino acids, proteins contain carbon, hydrogen, oxygen, nitrogen, and sometimes sulfur. One of the most important functions of proteins is their role as **enzymes** or catalysts. Enzymes accelerate chemical reactions in the body without additional energy, such as heat.

There are specific enzymes for specific reactions. The digestive system provides some excellent examples. The enzymes that react to digest starches are different from those that digest lipids in food. Because there are many thousands of different chemical reactions going on in the body at one time, there are many thousands of different enzymes.

Nucleic Acids

There are two types of nucleic acid—DNA, which stands for deoxyribonucleic acid, and RNA, which stands for ribonucleic acid. Nucleic acids are large molecules that are made up of nucleotides. Each nucleotide has four components: sugar, phosphate group, and a nitrogenous base. Here is more detail on each of these molecules (see Figure 1.1):

DNA

Sugar: deoxyribose

Bases: adenine, guanine, cytosine, or thymine

RNA

Sugar: ribose

Bases: adenine, guanine, cytosine, or uracil

As evident in the illustration, DNA looks similar to two ladders twisted together. These strands are the nucleotides, and their twisted form

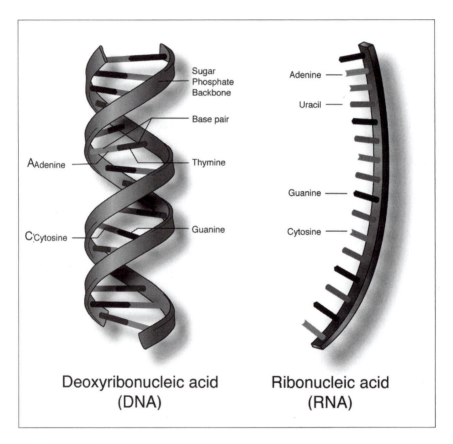

Figure 1.1 DNA and RNA Structure. A small portion of the RNA and DNA molecules, with some of the different nucleotides identified, including cytosine, adenine, and uracil. The synthesizing of RNA produces a complementary copy of half of the DNA strand. (National Human Genome Research Institute)

is called a double helix. Phosphate and sugar molecules alternate to form each nucleotide, or the upright length of the ladder. The rings that connect the strands are formed by the nitrogenous base pairs. It's important to note that in terms of the nitrogenous pairs, adenine is always with thymine, while guanine is always with cytosine.

DNA contains information on hereditary characteristics, and therefore is the body's genetic code. The code is determined in how the bases are arranged, and determines the various kinds of proteins produced by the body. While these bases make up the protein, the sequence of these bases is called a gene.

While DNA is a double strand of nucleotides, RNA is just a single strand. Uracil is in place of thymine in the nucleous bases. It is produced through synthesis from DNA in a cell's nucleus. RNA's primary function is to carry out protein synthesis, and it performs this duty in the cytoplasm.

Cells

The next level of complexity in the human body is **cells**, which are the smallest units in the body. There are numerous types of cells in the body, and different cells perform specific functions depending on their chemical reaction. However, in order for the body to be stable and function normally—also known as a state of **homeostasis**—cells must work together. While the specific functions of cells will be discussed throughout *The Encyclopedia of Human Body Systems*, it's important to have a basic understanding of the structure of a cell and some of their basic functions.

In addition to varying by function, cells also come in all shapes and sizes, although most are very small and can only be viewed under a microscope. They are measured in units called microns, and each micron is approximately 1/25,000 of an inch. Some exceptions to this are nerve cells, which can be very long, although very narrow in diameter. Some nerve cells are at least two feet long.

Despite their different shapes and measurements, most cells have four common structural features: a **cell membrane**, a **nucleus**, and **cytoplasm** and **cell organelles** (see Figure 1.2). The one exception is mature red blood cells—they have no nuclei (the plural form of nucleus).

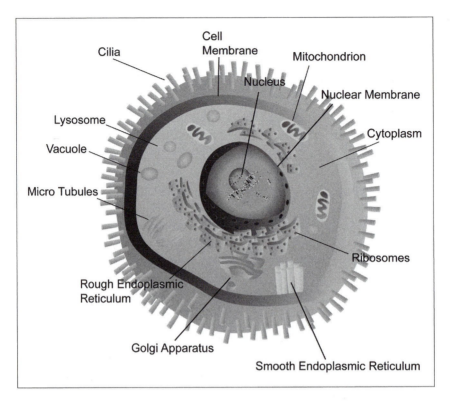

Cilia

Cell Membrane

Mitochondrion

Nucleus

Nuclear Membrane

Lysosome

Cytoplasm

Vacuole

Micro Tubules

Rough Endoplasmic Reticulum

Ribosomes

Golgi Apparatus

Smooth Endoplasmic Reticulum

Figure 1.2 Human Cell. Some important elements of a cell, including the nucleus, cytoplasm, mitochondrian, lysosome, and ribosomes. (Mark Rasmussen/Dreamstime.com)

Cell Membrane

The cell membrane is the outermost part of the cell, and is composed of phospholipids, cholesterol, and proteins. This membrane is permeable, which means that certain substances and fluids are allowed to enter and exit. In fact, the phospholipids allow other lipids or substances that are soluble in lipids to leave and enter the cell membrane through a process called diffusion. But the fluid movement into and out of the membrane is controlled by cholesterol in the cell, making the cell stable. It's important to note that while certain substances are allowed into the cell through its membrane, the membrane is selectively permeable. Part of the job of the membrane is to keep out dangerous substances that will harm the cell, and therefore the body.

Proteins are also involved in allowing substances to flow in and out of the membrane. Some cell proteins form openings known as pores that allow the stream of materials in both directions. Other proteins act as enzymes that enable substances to go in and out of the membrane. Finally, some proteins act as receptor sites for hormones. Once these hormones come into contact with these receptor sites, certain chemical reactions begin.

Nucleus

Located within the cytoplasm is the nucleus, which is protected by the two-layered nuclear membrane. Each nucleus has at least one nucleoli, as well as the cell's chromosomes. The nucleolus is made up of three substances: DNA, RNA, and protein. Chromosomes are made up of DNA and protein.

There are 46 chromosomes in a nucleus, and they are in the form of long threads called chromatin. As discussed earlier in this chapter, DNA is the body's genetic code. While the DNA in a nucleus does contain all of the body's genetic traits, there are only a few genes that are active. It is these active genes that code the proteins that determine the specific type of the cell.

Cytoplasm and Cell Organelles

Cellular chemical reactions take place in the cytoplasm, which is the watery liquid between the cell membrane and nucleus. Cytoplasm is a concoction of minerals, gases, and other organic molecules as well as cell organelles.

Each of these organelles has a specific job to do in order to enable cells to function. One type of cell organelle is the ribosomes, made up of protein and ribosomal RNA, that is the location of protein synthesis. Also present in the cytoplasm is the Golgi apparatus. These are flat, circular membranes arranged in a stack. Inside these sacs, materials such as carbohydrates are synthesized, and leave the cell by secretion. This secretion occurs when a single sac breaks from the Golgi membrane, becomes part of the cell membrane, and ultimately leaves the cell.

In addition to the ribosomes and the Golgi apparatus, another cell organelle is the mitochondria, which are oval in shape. The mitochondria have a double membrane, with the inner membrane containing folds known as cristae. Mitochondria are where cell respiration takes place, and the location of ATP, or energy, production.

Another type of organelle is lysosomes, which contain digestive enzymes and help white blood cells to destroy bacteria. These lysosomes also digest dead cells and damaged cellular parts, which has to occur before a cell can be repaired.

The last three types of organelles are centrioles, cilia, and flagella. Centrioles are located immediately outside the nucleus, and play an important role in organizing spindle fibers during the division of a cell. Cilia and flagella can be thought of as sweepers; they are responsible for sweeping certain materials along the surface of the cell.

Tissues

Following the cellular level, the next level is **tissues**, which is a cell group with a similar structure that performs a function. There are four types of tissue groups:

- Epithelial tissues: These cover or serve as lining on certain body surfaces, including the outer layer of the skin; some of these cells produce secretions that have a specific function.

- Connective tissues: These connect as well as support certain parts of the body. Some connective tissues transport and/or store materials. Examples include blood, bone, and **adipose** tissue.

- Muscle tissues: These are responsible for contraction, which enables movement. Examples include skeletal muscles as well as the heart.

- Nerve tissues: These are responsible for regulating body functions through generating and transmitting electrochemical impulses. Examples include the brain and optic nerves.

Organs

Groups of tissues arranged to perform specialized functions are called **organs**. Some examples include the stomach, kidneys, and lungs. The organs all work in concert with other each other. For example, the epithelial tissue in the stomach secretes juices that allow food to digest. The wall of the stomach contains muscle tissue that contracts, allowing food to mix with gastric juice. This digesting concoction then moves into the small intestine. Along the way, nerve tissue is transmitting an impulse that controls the contracting of the stomach.

Organ System

Just like organs are a group of tissues that perform a specific function, an organ system is a group of organs that together perform a specific function. Just like the organs work together, so do the organ systems to ensure the proper function of the human body. There are 11 organ systems, and each with their own chapter in this book. The body's organ systems are integumentary, skeletal, muscular, nervous, endocrine, circulatory, lymphatic, respiratory, digestive, urinary, and reproductive (see Table 1.2) for a more

TABLE 1.2
Organ Systems of the Human Body

Organ system	General function	Some examples
Circulatory	Movement of chemicals through the body	Heart
Digestive	Supply of nutrients to the body	Stomach, small intestine
Endocrine	Maintenance of internal environmental conditions	Thyroid
Integumentary	Protection against pathogens and chemicals, as well as excessive loss of water	Skin, hair
Lymphatic	Immune system, transport, return of fluids	Spleen
Muscular	Movement	Cardiac muscle, skeletal muscle
Nervous	Processing of incoming stimuli and coordination of activity	Brain, spinal cord
Reproductive	Production of offspring	Testes, ovaries
Respiratory	Gas exchange	Lungs
Skeletal	Support, storage of nutrients	Bones, ligaments
Urinary	Removal of waste products	Bladder, kidneys

detailed description of the functions of each organ system. It's important to understand that the organ systems work together in order for the body to function properly.

The Body's Cavities

Another important method to describe the human body is according to cavities. There are two major cavities in the body: the **dorsal cavity** and **ventral cavity** (see Figure 1.3). It's important to note that when describing

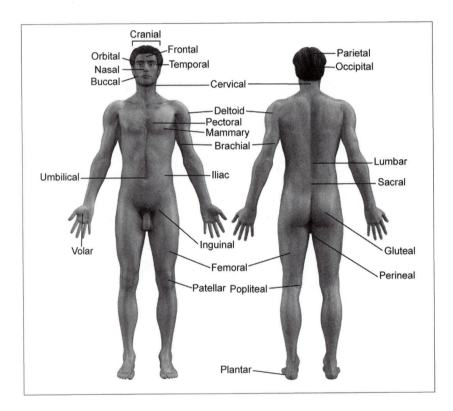

Figure 1.3   Body Parts and Areas. The anterior (left) and posterior (right) views of the human body, including body parts and areas. (Linda Bucklin/Dreamstime.com)

these and related locations, it's assumed that the body is in an anatomical position, which means that the body is standing up and facing forward. The arms are at the side and facing up.

Based on this anatomic position, the dorsal cavity is the posterior view. This cavity includes the cranial cavity, which is comprised of the skull and the brain. The dorsal cavity also includes the spinal cavity, which includes the spine and spinal cord. Both the cranial and spinal cavities contain all of the nervous system organs. Membranes known as **meninges** line the dorsal cavity systems.

The ventral cavity includes organs and organ system in the anterior view. The **thoracic** and **abdominal cavities** are part of the ventral cavity system. Located in the abdominal cavity is the **pelvic cavity**. The abdominal and thoracic cavities are divided by the diaphragm.

The thoracic cavity's organs include the heart and the lungs, which are covered by **pleural membranes**. The pleural membranes include the parietal, which cover the walls of the chest and visceral, which cover the lungs. Pericardial membranes cover the heart. Within the heart, there are parietal and visceral pericardium membranes that cover various areas of the heart. These membranes are known as **serous**, which means they cover organs.

In the abdominal cavity are located the liver, stomach, and intestines. There are two types of serous membranes in the abdominal cavity—the **peritoneum** and **mesentery**. The abdominal wall is lined by the peritoneum, while the outer surfaces of the organs located in the abdominal cavity are covered by the mesentery membranes.

The third cavity, the pelvic cavity, is below the abdominal cavity. The peritoneum is also present in this cavity, because it covers some of the surfaces of the pelvic organs, which include the body's reproductive organs, as well as the urinary organs such as the bladder.

Planes, Sections, and Areas

In addition to cavities, there are other important ways to describe how the body is organized. These include learning certain terms that refer to certain areas of the body (Table 1.3), as well as terms that refer to specific locations

TABLE 1.3
Key Anatomy Terms: Body Parts

Name of body part/area	Refers to
Axillary	Armpit
Brachial	Upper arm
Buccal	Mouth
Cardiac	Heart
Cervical	Neck
Cranial	Head
Cutaneous	Skin
Deltoid	Shoulder
Femoral	Thigh
Frontal	Forehead
Gastric	Stomach
Gluteal	Buttocks
Hepatic	Liver
Iliac	Hip
Lumbar	Small area of the lower back
Mammary	Breast
Nasal	Nose
Occipital	Back of the head
Orbital	Eye
Parietal	Crown of head
Patellar	Kneecap
Pectoral	Chest
Plantar	Sole of the foot
Pulmonary	Lungs
Renal	Kidney
Sacral	Spine base
Temporal	Side of head
Umbilical	Navel

in the body (Table 1.4). You might recognize some of these terms, such as visceral, which have already been mentioned in reference to cavities and membranes.

TABLE 1.4
Key Anatomy Terms: Location

Location/position in the body	Refers to
Superior	Above, higher
Inferior	Lower, below
Anterior	Towards the front of the body (chest)
Posterior	Towards the back of the body (lumbar)
Ventral	Towards the front of the body (breasts, mammary glands)
Dorsal	Towards the back of the body (buttocks)
Medial	Towards the body's midline (lungs)
Lateral	Facing away from the midline (shoulders, in comparison to neck)
Internal	Within (brain in the skull)
External	Outside (ribs are located outside of the lungs)
Superficial	Towards the body's surface (skin)
Deep	Within (various vein systems)
Central	Primary portion of the body (brain in relation to the nervous system)
Peripheral	Rooted in the main part of the body and extending outward (nerves in the arms, legs, and face)
Proximal	Near the origin (knee, in relation to foot)
Distal	Removed from the origin (bicep in relation to hand)
Parietal	Related to the cavity walls (membranes that line the chest cavity)
Visceral	Related to the organs within a cavity (membranes that cover the lungs)

One way the body is described—particularly the internal anatomy—is by separating different parts of the body into sections, planes, and areas (Planes are also discussed in Chapter 11, "The Skeletal System").

There are three primary planes in body: the **median sagittal plane**, the **transverse plane**, and the **frontal plane** (see Figure 1.4). The **plane** is an imaginary flat surface intersecting the body that separates the body into two separate sections of the body or of an organ, such as the brain or the spinal cord. It's important to note that the right and left refer to that of the person being described, not how it looks on the printed page. The median sagittal plane is longitudinal and divides the left and right parts along the

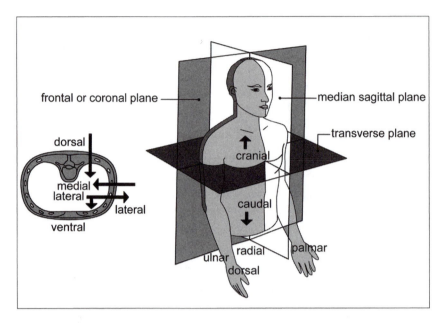

Figure 1.4   Planes of the Body. The left graphic represents a cross section of the spinal cord. The terms indicating directions of the body are indicated. The right graphic identifies the terms for the different planes of the body. (Sandy Windelspecht/Ricochet Productions)

midline of the body. The transverse plane divides a person into upper and lower sections, while the frontal plane divides the body in dorsal (posterior or lumbar side) and the ventral (anterior or abdominal side) portions.

Two other sections/planes are the cross section and the longitudinal sections. The cross section is a plane that runs perpendicular to the long portion of an organ. Examples include the cross section of the spinal cord or even a small intestine. A longitudinal section runs along the long axis of the organ.

Abdomen Areas

The abdomen—what is known as the lower trunk of the body—is divided into four quadrants, as well as nine areas that are based on the planes and sections that were just described (see Figure 1.5). These quadrants are

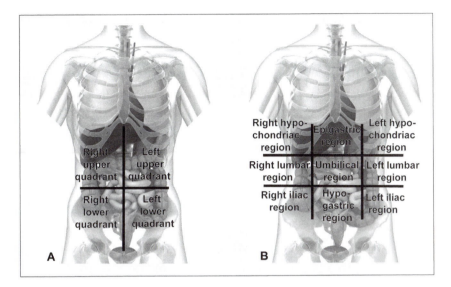

Figure 1.5   Areas of the Abdomen. The "A" graphic represents the four quadrants of the body. The "B" graphic depicts the abdomen's nine quadrants. (Sebastian Kaulitzki/Dreamstime.com)

evident when a transverse plane and a midsagittal plan intersect around the umbilical area. The four quadrants of the abdomen are the right upper quadrant, left upper quadrant, right lower quadrant, and left lower quadrant.

But to describe the different abdominal areas in more detail, the abdomen call be further divided into nine areas if two transverse planes and two sagittal plans divide the body. The nine regions can actually be organized according to three groups: the upper areas, middle areas, and lower areas. The upper areas include the hypochondriac and epigastric regions, while the middle areas include the lumbar and umbilical regions. The lower areas encompass the iliac and hypogastric regions.

Summary

Before understanding each of the body's complex organ systems, it's important to have an understanding of how the body is organized, as well as to get an overview of how the body's key building blocks function. This includes understanding that the body is divided into levels—beginning with the simplest structures like the chemical, cellular, and tissue levels, and moving up to more complex levels, including the organ, organ system, and organism levels. The body is made up of organic and inorganic chemicals, but its smallest unit is the cell. In addition to levels, the body is also organized according to cavities, as well as planes, sections, and areas.

2

The Circulatory System

Leslie Mertz

Interesting Facts

- At any given time, the veins and venules typically hold about two-thirds of the blood flowing through the body.

- As the heart contracts and blood rushes into the aorta, it is traveling at a speed of about 8 inches (20 centimeters) per second.

- Even in a person who is resting, blood issuing from the heart can travel down to the person's toes and back to the heart in just a minute. When a person is exercising heavily, that trip can take just 10 seconds. On average, every red blood cell completes the heart-to-body-to-lungs circuit 40–50 times an hour.

- If all of the blood vessels in an average adult were strung together end to end, they would reach at least 60,000 miles long, more than twice the distance around the Earth's equator. The capillaries alone make up 60 percent of that total.

- Every second, 10 million red blood cells die in the normal adult. The body replaces them just as quickly, however, so the total number remains constant.

- In the average adult, the heart weighs less than three quarters of a pound—about 11 ounces (310 grams). In any given person, it's about the size of his or her fist.

- The heart beats an average of 72 times a minute with a typical at-rest volume of 75 ml of blood pumped with each beat. Using those figures, a 75-year-old's heart has contracted more than 2.8 billion times and pumped more than 212 million liters of blood in his or her lifetime.

- When a person is resting, the left ventricle pumps about 4–7 liters of blood every minute. In a well-trained athlete who is doing strenuous exercise, that amount can rise to almost 30 liters per minute.

- Heart rate changes greatly during child development. The typical heart rate in a newborn is 130 beats per minute (bpm). It drops to 100 bpm by the time the child reaches 3 years old, 90 at 8 years old, and 85 at 12 years old.

- In an increasingly common practice, people are donating blood for use in their own upcoming surgeries. Called autologous blood donation, it helps patients ensure safe transfusions.

Chapter Highlights

- Blood: red blood cells, white blood cells, plasma
- Blood vessels: systemic and pulmonary circulation, arterial and venous systems
- Capillary system
- The heart: muscle, blood flow, electrical function
- Organs that depend on the circulatory system: digestion, liver and hepatic circulation, kidneys and renal circulation, spleen, cerebral circulation and the blood-brain barrier

Words to Watch For

Agglutination	Anions	Aortic arch
Albumins	Annulus fibrosus	Arterial baroreceptor
Amino acids	Antigens	reflex

Arterial system

Arteriovenous anastomoses

Atrioventricular node

Atrium

B lymphocytes (B cells)

Baroreceptor

Basophils

Bayliss myogenic response

Blood

Blood pressure

Blood sugar level

Blood type

Blood vessels

Bowman's capsule

Brachiocephalic artery

Bundle of His

Capillaries

Cations

Chordae tendineae

Circle of Willis

Circulatory system

Colic artery

Common carotid arteries

Complement

Concentration gradient

Coronary arteries

Coronary circulation

Diastole

Diffusion

End-diastolic volume

Endocardium

Endothelium

Eosinophils

Epitope

Femoral artery

Femoral vein

Fibrinogen

Fibular vein

Gastric

Gastroepiploic

Globulins

Glomerulus

Granulocytes

Heart

Heme group

Hemoglobin

Hemolysis

Hypertension

Immunoglobulins

In-series blood circulation

Intestinal villi

Involuntary muscle

Lymph

Lymphocytes

Macrophages

Microvilli

Monocytes

Nephrons

Neutrophils

Osmosis

Pacemaker

Partial pressure

Pericardium

pH level

Phagocytosis

Plasma

Plasminogen

Platelets

Portal circulation

Prothrombin

Protozoa

Purkinje fibers

Red blood cells

Septum

Sinoatrial node

Solutes

Stroke

Systemic circulation

Systole

T lymphocytes (T cells)

Terminal arterioles

Thrombocytes

Thromboplastin

Tunica adventitia

Tunica intima

Tunica media

Type A, B, AB, and
O blood

Ureter

Vasoconstrictor
nerves

Vena cava

Ventricle

Voluntary muscle

White blood cells

Introduction

Animals possess an array of internal circulatory systems for transporting materials through the blood and to every part of the body. Depending on the type of animal, the system may be quite simple or very complex. Earthworms, for example, have a series of "hearts" that are little more than pulsating blood vessels to assist the transit of blood through other vessels and to body organs. Insects and some other invertebrate animals have what is known as an open circulatory system that forgoes the network of vessels and instead usually delivers blood through one long, dorsal vessel that empties into a large cavity, or sinus. In that flooded cavity, the body organs are actually bathed in blood. As organisms become more complex and larger, a closed circulatory system is the norm. In a closed system, the blood makes its route through the body and to the tissues in vessels. Earthworms have a closed circulatory system, and so do all vertebrates, including humans.

The purpose of this chapter is to examine the structure and function of the human circulatory system, which is also known as the cardiovascular system because it includes the heart, or cardium from the Greek word for heart, and the blood vessels, or vasculature. The blood and all of its cells are also part of this system. The circulatory system serves as the body's delivery method, picking up oxygen from the lungs and dropping it off at tissue and organ cells around the body, then gathering carbon dioxide from the tissues and organs, and shipping it off to the lungs. It also distributes nutrients from the digestive system, transports chemical messages from the brain and other organs to various sites in the body, and provides the route and means for the body to mount a defense against bacterial infection. It even maintains the internal temperature by shuttling excess heat from the core of the body to the outside.

A Living River: Foundations of the Human Circulatory System

An amazing river system flows through the human body. After a person takes a breath of air, the oxygen is swept into the current and rushes to muscles, the brain, or another part of the body. Shortly after a Sunday dinner, nutrients begin to make their way into this same system for dispersal throughout the body. Alongside them, bacteria-fighting cells race to the site of an infection.

This remarkable river is the **circulatory system**: the **heart**, **blood vessels**, and **blood**. In a river system, water flows with a current between banks. A typical river system comprises tiny creeks, usually with a very slow current; larger, faster-moving river branches; and the main river with its strong flow. Likewise, the circulatory system has a liquid that flows with a current within a confined space. Blood replaces water. Instead of a river-bed, the blood flows inside of tubes, the blood vessels. Here and there, smaller branches separate from or connect into the main bloodstream. Blood from these branches eventually flows into or collects from even tinier vessels, which have a much slower current than the main bloodstream. This chapter will provide an introduction to the various components of the circulatory system, primarily the heart, the blood vessels, and the blood.

The Blood

Blood is more than just a simple, red liquid. It is actually a clear, somewhat gold-colored, protein-rich fluid crowded with **red** and **white cells**. The preponderance of red cells gives it the scarlet cast. When separated from the rest of the blood, the clear fluid, called **plasma**, has a more watery consistency. The reason that blood is more like syrup than water is the addition of red and white cells, and platelets, which combine to make up 40 to 45 percent of blood volume. Just as a glass of mud is more difficult to pour than a glass of water, because mud is actually a mixture of water plus dirt particles, blood is thicker because its plasma is laden with red and white blood cells. From this standpoint, blood truly is thicker than water. The "thickness" of a liquid is known as its viscosity. The slower something flows, the more viscous it is. Blood, for example, is three to four times more viscous than water.

All of the various components of blood have vital functions. As an example, the plasma serves as the liquid that suspends the red and white blood cells, along with all of the other chemical compounds and various materials that use the bloodstream to travel throughout the body. It also regulates the movement of heat from the body's core to the skin, the head, and the extremities. The red blood cells have a primary role of transporting oxygen from lungs to cells, while the white blood cells help defend against infection from invading organisms and foreign proteins. Table 2.1 describes some of these components in greater detail.

Red Blood Cells

Of the 5.5 quarts (5.2 liters) of blood in an average person, the red blood cells are, by far, the most prominent cellular component. Red blood cells, or **erythrocytes**, number about 28.6 trillion in the average male and 24.8 trillion in the average female. It follows, then, that red blood cells are microscopic. Ranging in size from 2.6 to 3.5×10^{-4} inches (6.5–8.8 m), red blood cells are disk-shaped cells with concave depressions in the centers of both sides. Red blood cells must be flexible, too. The flexibility is critical, because they have to bend, twist, and otherwise deform to squeeze through the tiny capillaries that serve as gateways to the tissue. Another identifying

TABLE 2.1
Components of the Blood

	Size	Lifespan	Number
Red blood cells[a] (Erythrocytes)	6.5–8.8 μm	120–180 days	5.5×10^{12}/L in males 4.8×10^{12}/L in females
White blood cells (Leukocytes)	7–18 μ	Variable	$4–11 \times 10^9$/L
Platelets	about 3 μ	4–10 days	$150–400 \times 10^9$/L

[a]The number of red blood cells increases among persons living at high altitudes. In extreme altitudes, individuals may have 50 percent more red blood cells than the amounts shown here.

feature of red blood cells is the lack of nuclei, a characteristic that sets them apart from other blood cells.

The primary duty of red blood cells is to transport oxygen and carbon dioxide. After a human breathes in oxygen, the red blood cells deliver it to the tissues. As tissue cells use the oxygen, carbon dioxide begins to accumulate. The red blood cells then pick up the carbon dioxide waste product and transport it back to the lungs, where it is discharged during exhalation. The jobs of picking up and delivering oxygen and carbon dioxide are accomplished through a large chemical compound known as **hemoglobin** (Figure 2.1). Located within red blood cells, hemoglobin also gives red blood cells their red color. The more oxygen the hemoglobin is carrying, the brighter red the blood. When the blood is carrying carbon dioxide rather than oxygen— when deoxygenated blood is returning from the tissues back to the lungs—blood takes on a dark maroon hue. The change in color is actually a result of a slight change in the three-dimensional configuration of hemoglobin when it is carrying oxygen.

Hemoglobin itself is a combination of a simple protein and an organic structure that contains iron ions. The protein, called a globin protein, contains four polypeptide chains, which are short strings of **amino acids**, the building blocks of proteins. In adults, the four chains normally come in two varieties: a pair of alpha chains and a pair of beta chains.

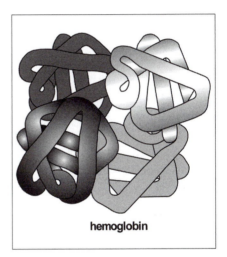

hemoglobin

Figure 2.1 A Hemoglobin Molecule. Hemoglobin is made up of four globin chains: two beta and two alpha chains. Within each globin chain sits one iron-containing heme group. The heme group binds oxygen for transport from the lungs to the body tissues. (Sandy Windelspecht/ Ricochet Productions)

Such hemoglobin is designated hemoglobin A. Each polypeptide chain is coupled with a separate iron ion, which is bound in a ringlike chemical structure known as a heme group. This heme group is the part of hemoglobin that actually binds oxygen for transport through the bloodstream.

Each red blood cell contains approximately 300 million molecules of hemoglobin, and every one of those units can bind with a total of four oxygen molecules—one oxygen for each of the four heme groups. Hemoglobin does not always bind oxygen, however. It has a differential ability to bind oxygen, which means that it picks up oxygen when the oxygen content in surrounding tissues is high, as it is in the lungs, and drops off oxygen when the oxygen content in the surrounding tissues is low, as it is in the tissues. This relative oxygen concentration is known as **partial pressure**. This property makes hemoglobin an ideal oxygen transportation vehicle. The high partial pressure existing in the lungs stimulates hemoglobin to load up with oxygen, and the low partial pressure in tissues triggers hemoglobin to release it. Sometimes, particularly under conditions of high acidity in the blood or elevated temperature, the red blood cells' affinity for oxygen can drop. This can cause oxygen delivery to tissues to similarly drop. (The relationship between pH and oxygen is discussed further in the section on plasma.)

The blood is also an excellent carrier for carbon dioxide, a by-product of cell metabolism. As carbon dioxide enters the red blood cells at a tissue site, it lowers the hemoglobin's affinity for oxygen, which further facilitates the discharge of oxygen into the tissues. Once in the blood, the carbon dioxide mostly travels either bound to hemoglobin or as bicarbonate ions (HCO_3^-) that form when carbon dioxide is hydrated (combined with water). The majority of CO_2 takes the latter form. Once the blood arrives at the lung, the bicarbonate ions revert to their original CO_2 state and depart through the lungs. The hemoglobin drops off its carbon dioxide for a similar exit, and the hemoglobin is ready to accept oxygen once again.

Besides oxygen, hemoglobin can lock onto dangerous gases like carbon monoxide (CO), a molecule that contains only one oxygen molecule, monoxide (carbon dioxide has two). Carbon monoxide is hazardous to human health because hemoglobin binds with carbon monoxide molecules 200 times more readily than it does with oxygen molecules. Therefore, it preferentially binds carbon monoxide instead of oxygen, which can severely reduce oxygen flow to tissues and can quickly become fatal. Cigarette smoke, as well as emissions from automobiles and many home heating systems, contains carbon monoxide. The potential for **carbon monoxide poisoning** from this gas, which is colorless and odorless, has prompted health professionals to warn people against running a car in a closed garage and to recommend the use of carbon monoxide detectors to check for a buildup of the gas in a heated home.

White Blood Cells

White blood cells have a completely different function than red blood cells. White blood cells, or **leukocytes**, are part of the body's defense team and can actually move out of the bloodstream to do their work in the tissues (Figure 2.2). Adults may have anywhere from 20 to 60 billion white cells

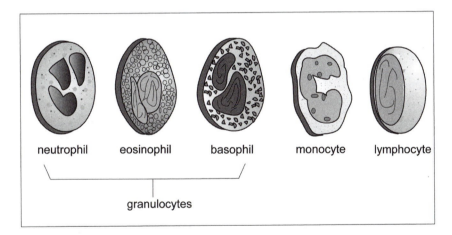

Figure 2.2 White Blood Cells. White blood cells come in three main types: granulocytes, monocytes, and lymphocytes. The granulocytes, so called because of their grainy appearance, include the neutrophils, eosinophils, and basophils. White blood cells are part of the body's defense system against foreign materials and invading microorganisms. (Sandy Windelspecht/Ricochet Productions)

in the bloodstream, far fewer than the 25 trillion red blood cells in the human body. White blood cells come in three main types:

- Granulocytes, including neutrophils, eosinophils, and basophils

- Monocytes

- Lymphocytes

Granulocytes are the most abundant type of white blood cell, comprising 7 out of every 10 leukocytes. They are named granulocytes based on the grainy appearance of the cytoplasm, the part of the cell outside the nucleus but inside the membrane. All granulocytes also have distinctive lobed nuclei. Depending on which type of biochemical dye best stains them, granulocytes are further subdivided into **neutrophils, eosinophils**, or **basophils**. Neutrophils stain with neutral dyes, basophils stain with basic dyes, and eosinophils readily stain with the acid dye called eosin.

In addition to their different staining properties, the three types of granulocytes have separate functions. Neutrophils, the most common granulocyte with up to about 5.2 billion cells per quart (5 billion cells per liter) of blood, engulf and destroy small invading organisms and materials, which are collectively known as **antigens**. Averaging about twice the size of red blood cells, neutrophils are a main bodily defense mechanism against infection and are particularly suited to consuming bacteria. This process of engulfing and destroying bacteria and other antigens is called **phagocytosis** (Figure 2.3). The sequence of events begins with the human body recognizing that a foreign material has invaded. Antigens are different from the body's own cells and trigger the body to enter its defense mode. If the infected site is within the bloodstream, the neutrophils remain there, but if the site is in the tissues, the neutrophils will flow out through the capillaries to flood the area of infection directly. Each neutrophil at the infected site stretches a bit of its tissue, called a pseudopod, toward and then around the invader. Once the invader is contained inside the neutrophil, an organelle called a lysosome finishes the job by using its internal battery of enzymes and hydrogen peroxide to digest the material. Usually, the neutrophils are able to kill the bacteria quickly, but sometimes the toxins in the bacteria are

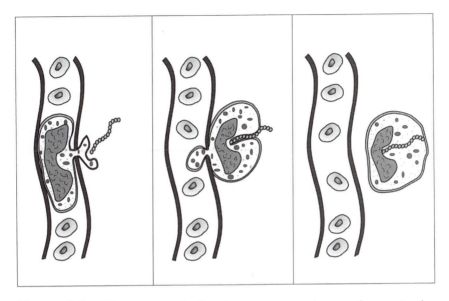

Figure 2.3 Phagocytosis. Leukocytes are primary players in the body's defense mechanism. Here, a leukocyte protrudes from a blood vessel and surrounds an invading bacterium. Once the bacterium is engulfed, it is destroyed. This process of engulfing and destroying materials is called phagocytosis. (Sandy Windelspecht/Ricochet Productions)

fatal to the neutrophil. The result is pus, a mixture of mostly dead bacteria and leukocytes that perished during the battle.

Basophils, which are the smallest and least common type of white blood cells, appear to be active in the inflammatory process. While basophilic activity is not fully understood, scientists do know that basophils release substances like histamine and serotonin. Histamine helps maintain a free flow of blood to inflamed tissues, particularly by dilating blood vessels. Serotonin is similarly vasoactive. Basophil granules also contain heparin, which helps prevent blood from clotting.

The third group of granulocytes, the eosinophils, likewise demands additional study. Evidence indicates that they engage in minimal phagocytosis. Their primary tasks are to moderate allergic responses and to help destroy parasites. They accomplish the latter by using unique proteins that are toxic to specific invading organisms. Although they are not quite as motile as the basophils and neutrophils, eosinophils do slowly travel through the body.

The second major type of white blood cell is the **monocyte**, the largest cell in the bloodstream. Monocytes are much larger than red blood cells, and have diameters of $3.9–11.8 \times 10^{-4}$ inches (10–30 m). They are, however, only temporary residents in the bloodstream, remaining in the blood for about three days before moving into tissues where they become macrophages, large cells that engage in phagocytosis. Just as the neutrophils do, the highly motile monocytes latch onto invading organisms, then literally devour them with a mixture of highly reactive molecules. The neutrophils typically target smaller organisms, like bacteria, while the macrophages take on larger invaders, even **protozoa**, and also remove old cells and other detritus from the bloodstream.

Lymphocytes, the third major type of leukocyte, are the second most common white blood cell, but they frequently are not in the bloodstream, either. They usually reside in the **lymph**, a clear, yellowish fluid that exists around and between cells in the body tissues. This fluid, which is about 95 percent water, enters the bloodstream mainly through one of two ducts, and carries lymphocytes into the blood with it. Lymphocytes come in two main varieties: **B lymphocytes**, or **B cells**, and **T lymphocytes**, or **T cells**. Both B and T cells have antigen receptors on their cell surfaces. These receptors are highly specific. In other words, one particular form of lymphocyte can only bind to one type of foreign material, much as different keys fit different locks. The specific area of an antigen to which the B cell receptor binds is called an **epitope**. (For more information on the functions of B cells and T cells, see Chapter 6 on the lymphatic system.)

Plasma

The circulatory system reaches just about everywhere in the human body, so the volume of plasma is fairly high. In fact, plasma makes up about 5 percent of a normal human's body weight. Plasma itself is a solution of about 90 percent water, 7–9 percent proteins, and roughly 1 percent ions, which are either positively or negatively, charged molecules of such chemicals as sodium, calcium, and potassium. The remainder includes dissolved organic nutrients, gases, and waste products. Some textbooks refer to the plasma as the extracellular matrix of blood. In other words, it is the portion of blood that lies outside the red and white blood cells, and that provides the blood's liquid "structure."

Plasma is mostly water, but it is the smaller ion and protein components that draw biologists' interest. Their roles include regulating the blood's volume and viscosity; maintaining the steady **pH level** required by the muscular, nervous, and other major physiological systems; facilitating the transport of various materials; and assisting in bodily defense mechanisms, including the immune response and blood clotting.

Plasma proteins consist of **albumins** (the most abundant plasma protein), **globulins**, and **fibrinogen**, most of which are synthesized in the liver. The ions in plasma include both positively and negatively charged varieties. Sodium (Na^+), calcium (Ca^{2+}), potassium (K^+), and magnesium (Mg^{2+}) carry positive charges and are called **cations** (pronounced "cat-ions"). **Anions** ("ann-ions") are negatively charged, and in the plasma include chloride (Cl^-), bicarbonate (HCO_3^-), phosphate (HPO_4^{2-} and $H_2PO_4^-$), and sulfate (SO_4^{2-}). Most of the ions in the plasma of humans and other mammals are sodium and chloride. The general public is familiar with the two ions in their combined form of NaCl, or normal table salt.

Because blood vessels are permeable to water, water can move freely into and out of the blood. The more water in the blood, the greater the overall blood volume. A change in the concentration of different ions or proteins can play havoc with the amount of water in the blood, and too much or too little water can have damaging effects on a person's health. The body's ion and protein concentrations keep the blood volume from plummeting too low or rising too high. Plasma proteins and ions make up 10 percent or less of the plasma volume, but they are critical in regulating how watery the plasma, and thus the blood, is. Sodium chloride (NaCl) and sodium bicarbonate ($NaHCO_3$) are the key ion regulators of the amount of water in the plasma, while the albumins are the primary proteins involved in the water content of the plasma. This system is based on a balance between the liquid inside the blood vessels and the liquid outside.

Osmosis is a process that seeks to equalize the water-to-solute ratio on each side of a water permeable membrane. In other words, if the water on one side of the membrane has twice as many dissolved materials, called **solutes**, additional water will move by osmosis across the membrane and into the side with the higher solute concentration. This action adjusts the solute concentration so that the ratio of water to solutes on each side of the membrane is the same. In blood vessels, water likewise moves in and out based on the relative solute concentrations within and beyond the vessels. As it turns out, the water content in the circulatory system sometimes falls too low, because the sheer force of blood rushing away from the heart literally pushes water out of the smallest blood vessels, the capillaries. The system remains stable because plasma proteins and the albumin proteins, in particular, diffuse very poorly through capillary walls and therefore allow the osmotic pressure gradient to draw water back into the capillaries.

Osmosis alone is not enough to maintain the proper solute concentration in blood. Blood carries all kinds of solutes, including organic molecules like food, cholesterol and other fats, waste products, and hormones, yet the blood is not continually flooded with water. The reason is that ions, specifically sodium, use pumps that reside in the membranes of the cells within the blood vessel to actively drive sodium molecules out of the vessels, leaving behind a lower concentration of solutes and, in turn, requiring less water to enter by osmosis. This balance between osmosis and ion pumps is vital to regulating blood volume.

While the total amount of ions and proteins is important, the levels of individual ions and proteins are also significant. For example, muscles and nerves respond to even slight changes in potassium and calcium ion concentrations. Likewise, cell membranes rely on the right combination of calcium, magnesium, potassium, and sodium in their immediate environments.

As mentioned, plasma proteins include albumins, globulins, and fibrinogen. Fibrinogen makes up only 0.2 percent of plasma proteins, with the remaining 99.8 percent split as 55 percent albumins and 44.8 percent globulins. As previously described, the albumins are involved in maintaining blood volume and water concentration, and some of the globulins serve as transportation vehicles for a variety of molecules. Other important activities for globulins include blood clotting and immune responses. The immune response of plasma proteins will be discussed here, and blood clotting in the next section. (More details are available in Chapter 6 on the lymphatic system.)

Globulins come in three types: alpha, beta, and gamma globulins. The transferrin used in iron transport is a beta globulin. A variety of other beta globulins is collectively termed **complement**, and assist in the immune system by binding to passing potential antigens. Beta globulins identify invaders by telltale structures on the cell surface that are different than those of the body's own cells. Beta globulins are designed to grab onto these unusual structures, and they basically put a plug in the antigen's active site that renders it harmless. Similarly, beta globulins can ferret out cells that have antibodies already attached. In this case, another part of the body's immune response has already begun to mount a defense to the invading organism or foreign protein by creating the antibody. The beta globulins recognize the antibodies and attach to them. The captured cell then proceeds to the white blood cells for destruction.

Some plasma proteins, called **immunoglobulins** (Ig), go a step further and act as antibodies themselves. The five main types of these immunoglobulin antibodies are:

- IgA, which is found in bodily secretions, like saliva, tears, milk, and mucosal secretions

- IgE, which causes the sniffing and sneezing associated with hay fever and asthma, and also defends against parasites

- IgG, which helps battle infections and also confers mother-to-fetus immunity

- IgM, common to almost every early immune response

- IgD, which has an unknown function

Proteins within each type of immunoglobulin are similar in basic structure, but have a variable region specific to a particular antigen. (Details are available in Chapter 6 on the lymphatic system.) Another of the body's lines of defense depends in part on plasma proteins. Blood clotting requires fibrinogen, a soluble plasma protein; the beta globulins called **prothrombin** and **plasminogen**; platelets; and a slew of other molecules. The process is discussed in the next section.

Platelets and Blood Coagulation

Platelets, plasma proteins, vitamin K, and calcium all take their place in a quick-acting series of chemical reactions that result in blood coagulation, or clotting. Clotting begins almost immediately after the wound occurs as platelets congregate at the site of the injury. Platelets, also known as **thrombocytes**, are not cells. Rather, they are sticky, disk-shaped fragments of large blood cells called megakaryocytes that reside solely in the bone marrow. These small cell fragments exist throughout the circulatory system, with tens of millions in every droplet of blood. Their primary role is blood-clot formation. Because so many exist in the blood, a good supply of platelets usually is not far from the wound site.

The first step in blood clotting is the release by the damaged tissue of a substance known as **thromboplastin**. As platelets arrive at the wound site, they disintegrate and release additional thromboplastin. Thromboplastin and calcium are both required to trigger the beta globulin called prothrombin to produce the enzyme thrombin. For the next step, the thrombin, platelets, and fibrinogen, a soluble plasma protein, work together to help make a tight web of insoluble fibrin threads that stick together and to the blood vessel wall. When blood cells encounter the web, they become trapped and form a blood clot. A scab is a dry, external clot. A bruise is a blood clot, too, but an internal one.

Although the coagulation process may seem complex, it happens very quickly. Small cuts are usually sealed within a couple of minutes, with an external scab hardening in place not long afterward. The yellowish fluid sometimes remaining at the injury site is called serum. Although the term serum is sometimes used interchangeably with plasma, serum actually refers to plasma that no longer contains fibrinogen or other clotting factors. Normally, blood clots promote injury recovery by stopping blood loss, but that is not always the case. Clots that form within the blood vessels can be dangerous, because they can block blood flow and oxygen transport. A **stroke**, for example, is the result of a blood clot in the brain. Fortunately, platelets normally do not stick to the smooth walls of healthy, undamaged vessels. Other fail-safes are heparin, which is found in basophils, and substances called antithrombins that turn off thrombin activity, effectively shutting down the coagulation machinery and preventing unnecessary blood clotting.

Blood Type

The preceding introduction to blood cells may give the impression that blood in all individuals is alike. It is not. The most obvious differences are blood type and Rh factor: Human blood types, or groups, are **type A**, **type B**, **type AB**, or **type O**, and Rh factors are defined as either positive or negative.

Blood type refers to the presence or absence of chemical molecules on red blood cells. These molecules can instigate antibody reactions and are therefore antigens. Red blood cells can have one, both, or neither of the two antigens named "A" and "B." Blood with only A antigens or only B antigens is called type A or type B, respectively. Blood with both A and B antigens is called type AB, and blood with neither is type O. People with type A blood are also born with beta (or anti-B) antibodies, which are designed to detect and eliminate B antigens. Likewise, people with type B blood have alpha (or anti-A) antibodies that assail A antigens. Type AB blood has both antigens but neither antibody, and type O blood has neither antigen but both antibodies.

This confusion of letters means that type A blood donors can safely give their blood to any person who does not have antibodies to the antigens in their blood, namely A. As noted in Table 2.2, the type A donor's blood is compatible with the blood of recipients with type A or type AB. On the other hand, a person who has type A blood can receive blood donations from any person whose blood does not trigger a response from their own antibody contingent, beta. The table shows that type A persons can receive blood donations of type A and type O, because neither adversely reacts with the beta antibody in type A blood.

TABLE 2.2
Blood Types

Blood type	Antigens present	Antibodies present	Can be donated to	Can accept donations from
A	A	Beta	Type A, type AB	Type A, type O
B	B	Alpha	Type B, type AB	Type B, type O
AB	A, B	None	Type AB	Type AB, type A, Type B, type O
O	Neither	Alpha, beta	Type O, type A, Type B, type AB	Type O

Blood type refers to the presence or absence of chemical molecules on red blood cells. These molecules, called antigens, can instigate antibody reactions. For this reason, medical professionals check blood compatibility before performing transfusions.

Reactions between mismatched blood can be severe. If type A blood from one person is given to another person with type B blood, the blood will clump due to a process called **agglutination**, as the alpha antibodies battle the B antigen. After clumping, the red blood cells will rupture in a process called **hemolysis**, which can lead to serious consequences, such as kidney dysfunction, chills, fever, and even death. For this reason, medical professionals compare blood type and Rh factor from a patient and a donor before proceeding with a transfusion.

Type AB-positive blood is frequently called the "universal recipient." Type AB blood has neither alpha nor beta antibodies, which means that any blood can be introduced without the chance of an antibody attack by the recipient's blood. The recipient does not have to worry about antibodies from the donor blood, because the amount of donated blood is small, becomes diluted in the recipient's blood, and presents no threat.

At the opposite end of the spectrum, type O blood is known as the "universal donor." It has neither the A nor B antigens, and therefore can be administered with little fear of agglutination. Nonetheless, medical professionals still take precautions to ensure blood-transfusion compatibility by mixing donor and recipient blood and watching it closely for adverse reactions. The reason for the wariness is that A and B are not the only antigens. Sometimes, blood contains less common antigens that can bring about agglutination and cause problems for the patient.

Blood Vessels: The Transportation System Within

As previously described, the blood is not just red liquid. It is filled with millions of cells, each with a specific job to do. Likewise, blood vessels are much more than a simple series of pipes to contain and route blood. Blood vessels are living, dynamic tissues with complexities all their own.

The circulatory system in a single adult human being comprises some 60,000 miles of blood vessels. Most people can see at least a few of them just under the skin of the wrist. The vast majority is much smaller than those visible vessels, and has diameters of less than one three-thousandths of an inch (about 10 microns). Blood vessels are associated with one of three major groups: the arterial system, the venous system, or the capillary system.

Whenever blood is moving away from the heart, either to the body tissues (systemic circulation) or to the lungs (pulmonary circulation), the arterial system is involved. The capillaries take over when the blood reaches its destination, and serve as the exchange vessels between the blood and the lungs, or between the blood and the body tissues. When the exchange is complete, the blood moves from the capillaries into the vessels of the venous system, which directs the blood back to the heart to begin another route either to and from the lungs, or to and from the body tissues.

All three types of vessels, then, participate in the circuitous path of the blood from the heart to the lungs and back to the heart, and from the heart to the other body tissues and back. The heart-to-lungs-to-heart path is called the **pulmonary circulation** and serves to allow the blood to pick up oxygen. The heart-to-body tissues-to-heart circuit is called the **systemic circulation** and allows tissues to take up oxygen and other materials transported in the blood. Overall, the blood travels more or less in two loops, one from the heart to the lungs and back, and a second from the heart to the body tissues and back to the heart.

Arterial System

The main function of the **arterial system** is to carry blood away from the heart and either to the lungs to pick up oxygen, or to other body tissues to drop off nutrients, oxygen, hormones, or other needed substances. Like a river system that has main branches from which diverge many smaller side creeks, the arterial system has main branches called arteries and many diverging, smaller vessels called arterioles (Figure 2.4). The major arteries provide quick, direct routes to major body areas, and smaller arteries divert blood to more specific sites. Separating from the arteries, the arterioles bring blood to specific target tissues. The arterial system works much as a road system does: travelers use superhighways to get quickly to a general region, then take smaller freeways and finally side roads to reach a specific destination. In the case of the arterial system, the blood moves along main arteries, then into smaller arteries and even smaller arterioles. The specific destination of the blood cells is a set of capillaries in the lungs or in some other body tissue. For more information on a dangerous condition associated with the arteries, please read Sidebar 2.1.

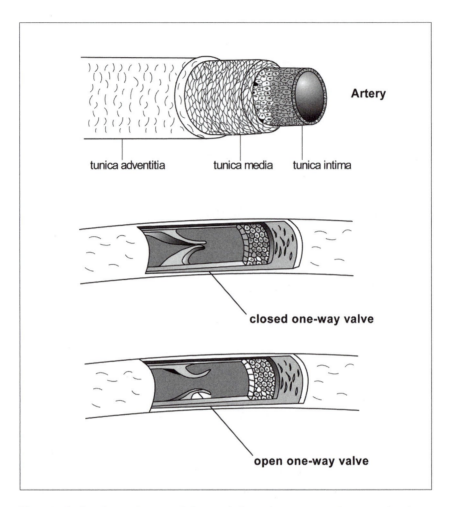

Figure 2.4 Arteries and Arterioles. Arteries and arterioles have three layers. The outer layer, called the tunica adventitia, is made of fibrous connective tissue and loosely holds the vessel in place. The middle, or tunica media, is a thick, muscular, elastic layer. The inner layer, or tunica intima, is a one-cell-thick layer of endothelial cells. Many of the larger blood vessels contain valves that open when blood is pulsing forward and close against any back flow. This ensures a unidirectional blood flow. (Sandy Windelspecht/Ricochet Productions)

SIDEBAR 2.1
Reducing the Risk Factors for Coronary Artery Disease

The leading cause of death in the United States is coronary artery disease (CAD), according to the National Heart, Lung, and Blood Institute. It is estimated that every year, over 500,000 Americans die from this disease.

CAD, also known as coronary heart disease, occurs when plaque builds up inside the arteries that supply the muscles of the heart with blood. This plaque buildup makes it harder for blood to get to the right areas of the heart. Plaque is composed of substances found in the blood, like fat and cholesterol. Plaque buildup results in a condition called **atherosclerosis**.

Certain factors increase the risk of developing CAD:

Overweight or obesity: Extra body weight from bone, fat, and/or water (overweight), or a high amount of extra body fat (obesity).

Unhealthy cholesterol levels: High LDL cholesterol (also known as bad cholesterol) and/or low HDL cholesterol (good cholesterol).

High blood pressure: Blood pressure that stays at or above 140/ 90 mmHg over a period of time is considered too high.

Smoking: Smoking can cause numerous risk factors. It can damage and tighten blood vessels, increase cholesterol levels, and inhibit oxygen from reaching certain tissues in the body.

Diabetes: In people with diabetes, the body's high blood sugar level means that it cannot use the hormone insulin properly or produce enough insulin.

A sedentary lifestyle: Lack of physical activity can exacerbate these and other risk factors, such as obesity and overweight.

Age: The risk of CAD increases with age. For men, this risk increases after age 45, and for women, the risk increases after age 55. During middle age, signs or symptoms of CAD can begin to present, particularly if there are genetic and lifestyle risk factors.

Genetic risk factors: Having a family history of CAD increases the risk of CAD.

Vessel Composition

Arteries and arterioles are more complicated than they might seem at first, and that complexity begins with the structure of the vessels themselves. Arteries and arterioles are made of three concentric layers: the **tunica adventitia**, the **tunica media**, and the **tunica intima**. The tunica adventitia coats the outside of an arterial vessel and is made of fibrous connective tissue that loosely holds the vessel in place. In the larger arteries, the adventitia also holds a number of small blood vessels of its own. These small vessels, called **vaso vasorum**, provide nourishment to the thick vessel walls.

The thickest of the three layers is the tunica media, which is the muscular and elastic middle layer of vessel walls. This layer allows the large arteries to expand and contract in tune with the waves of blood accompanying each beat of the heart. Widening slows the blood flow and narrowing quickens it, so the combined action helps to moderate the blood's speed. Elastin, a protein that has six times the spring of rubber, provides the vessels' elasticity and allows the vessels to stretch wide as a pulse of blood arrives. A much stiffer protein, called collagen, prevents the vessels from expanding too much. Stretching is essential for arteries lying close to the heart, where the force of the heart's pumping on blood flow is most strongly felt. Here, arteries distend to take the brunt of each blood rush, then recoil to create a more even blood flow to subsequent areas of the circulatory system. As might be imagined, these arteries contain a higher percentage of elastin than other vessels further down the line in the systemic circulation.

For the contraction of arteries, the spindle-shaped, smooth muscle cells of the tunica media take over. Overall, the body has three types of muscle: smooth, striated, and cardiac. Striated muscles are those that a person can consciously control. Leg muscles are an example. A jogger or walker can control the muscles to speed the pace or to slow down. Smooth muscle, on the other hand, is controlled mainly by the autonomic nervous system, which means that these muscles work involuntarily or outside of a person's will. Unlike striated muscle that tires rather quickly (as any jogger will attest), smooth muscle can continue working for long periods of time. The tunica media, or middle layer of the vessels, has smooth muscle, which wraps around the vessel rather than running its length and is responsible for contracting arteries.

When the smooth muscle tightens and decreases a vessel's diameter, blood pressure rises because the blood is forced through a narrower opening. The effect is similar to that achieved by holding the thumb partially over the water flow at the end of a garden hose. The decreased diameter of the hose at the point of constriction (the thumb) increases the water pressure. As stated earlier, the arteries closest to the heart have the most elastin to slow the strong pulses of blood leaving the heart. They also have the least smooth muscle. In contrast, vessels farther away from the heart have a higher proportion of smooth muscle to control vessel diameter and help keep the blood moving. The innermost layer of the blood vessel is the tunica intima, which is also known as the **endothelium**. This layer has direct contact with the blood that runs through the artery or arteriole. Although it is only one cell thick, the tunica intima's flat and smooth cells are important in imposing a barrier of sorts and preventing the passage of plasma proteins out of the blood. In arteries that connect with the heart, the tunica intima has a thin, fibrous layer that blends seamlessly with the heart's inner lining, known as the **endocardium**.

Systemic Circulation

The blood entering the **aorta** from the heart has just been to the lungs, so it is fully oxygenated and bright red. As just described, the aorta is highly elastic and can distend greatly to accept the powerful rush of blood pumped from the heart. As the aorta returns to its original size, the blood moves out in a more even flow. Although the blood eventually flows very smoothly, the pulse can be felt by pressing on some areas of the body where main arteries run close to the skin surface. For example, nurses typically determine a patient's heart rate by feeling the wrist, or the radial pulse site, and counting the number of pulses over a set period of time. A normal resting adult's heart rate is about 70 beats per minute. The rate is typically lower in adult athletes and higher in children. Other major arterial pulse sites include:

- Temporal, in front and slightly above the ear

- Facial, along the lower jaw

- Carotid, beside the windpipe in the neck

- Brachial, on the inside of the elbow

- Femoral, on the upper thigh beside the groin

- Popliteal, on the back of the knee

- Posterial tibial, on the inner ankle

- Dorsal pedal, just above the toes on the upper foot

The aorta is a large, arched artery that originates at the heart, where it connects to the lower left heart chamber, called the **ventricle**. From there, it continues down the trunk of the body. All of the major arteries in the human body branch off of the aorta. Using the analogy of a road system, the aorta would be the main superhighway through which all outgoing (systemic) traffic has to pass. The highways that divert from it would represent the major arteries. These major arteries supply blood to all of the main body regions, including the limbs, head, and body organs. The arterioles are the side roads that bring the blood to specific locations throughout the body.

Two major arteries parting from the **aortic arch** are the right and left **coronary arteries**, which provide blood to an important body organ, the heart. These two arteries further subdivide. The left coronary artery splits into a circumflex branch that runs behind the heart and an anterior descending branch that curves forward over the heart. These two branches mainly supply the left ventricle and left **atrium**, which is the heart's left, upper chamber. The right coronary artery and its posterior descending branch deliver blood mainly to the right side of the heart, which has its own ventricle and atrium. In addition, both the left and right coronary arteries supply blood to the **septum**. The septum is the interior wall between the right and left ventricles.

All of the blood pumped from the heart into the aorta in one contraction is called the **total cardiac output** or **stroke volume**. Because the heart is such a hard-working organ, it receives a rather large portion, about 5 percent, of the total cardiac output even when a person is resting. As will be discussed later, that proportion can change if a person is active or under some form of psychological or physical stress.

Blood travels to the head via two major vessels called the left and right common carotid arteries. The left carotid splits directly from the aortic arch between the bases of the two coronary arteries. The right carotid indirectly branches from the aorta via a short vessel, called the brachiocephalic (or innominate) artery. The brachiocephalic artery also feeds the right subclavian artery, which is discussed below. Each of the two carotids splits into internal and external carotid arteries, which are the principal arterial vessels of the head and neck.

Other arteries that branch from the aorta soon after it leaves the heart include the right and left subclavian arteries. Like the right carotid artery, the right subclavian artery divides off of the brachiocephalic artery. Each subclavian artery supplies an arm. They earned the name subclavian because their paths to the arms run beneath the collarbone, or clavicle. Blood flow continues down the length of the arm through the brachial artery and into the ulnar arteries and radial arteries of the forearm.

As the aorta travels down the spine, it is called the thoracic aorta in the chest, or thorax, and then becomes the abdominal aorta. Along the way, it supplies blood to various internal organs, such as the kidneys, spleen, and intestines. Each of the two kidneys, for example, gets its arterial blood supply from a renal artery that separates from the aorta. The renal artery splits into smaller and smaller vessels, eventually leading to a round cluster of capillaries, which is called the glomerulus. Fully 25 percent of the total cardiac output goes through the renal arteries to the pair of kidneys. These two organs not only excrete waste products through the urine but also have important roles in regulating the amount of water in the blood, as well as its pH level (a measure of acidity/alkalinity) and electrolyte content.

Some of the other major organs and systems receiving blood from the aorta—in this case, the abdominal artery—are the spleen, which is fed by the splenic artery; the liver, which gets its blood via the hepatic artery; and the digestive system, which is supplied by a number of arteries, including the gastric, gastroepiploic, superior and inferior mesenteric, sigmoidal, and colic arteries.

Many of the arteries' names come from the medical names of their destination. The renal artery is named for the renal, or kidney, system; the ulnar and radial arteries ship blood to the area surrounding the ulna and radius, which are bones in the forearm; and the superior mesenteric artery supplies the region around the intestinal membrane, which is known as the mesentery.

The abdominal artery ends when it bifurcates into the left and right common iliac arteries, each of which soon divides again into internal and external iliac arteries (sometimes called hypogastric arteries). From each of the internal iliac arteries come various arteries that supply blood to the pelvic area, including the reproductive organs. The external iliac artery becomes the femoral artery when it enters the thigh.

Continuing with the leg, the femoral artery of the thigh becomes the popliteal artery at the knee (popliteal is a medical term for the back of the knee), then divides into the posterior and anterior tibial arteries (the tibia is a bone in the lower leg).

As described previously, the arterioles-to-capillaries-to-venules route is the most common pathway for blood. However, the arterioles in a few tissues never connect with capillaries, instead attaching to venules by way of wide vessels called **arteriovenous anastomoses**. These muscular vessels, which range from 7.9×10^{-4} to 5.3×10^{-3} inches (20–135 µm) in diameter, are common in the skin and in the nasal mucosa (in the nostrils), and regulate body temperature. Heat from the body's core is transported via the blood to these areas for release to the outside.

Pulmonary Circulation

Pulmonary circulation is less complex than systemic circulation because the only target organ for the arterial system is the lungs. There, the blood picks up the oxygen that it will eventually deliver to the body through the systemic circulation, which was just discussed.

Unlike the systemic arterial circulation, which heralds from the heart's left ventricle at the aorta, the pulmonary arterial circulation begins with the right side of the heart and the pulmonary artery. The pulmonary artery connects to the right ventricle, which is the smaller of the heart's two ventricles. As the right ventricle contracts, it pushes blood into the pulmonary artery. The artery soon splits, and its two branches lead to either the right or left lung.

Just as blood in the systemic circulation moves from major to smaller arteries, and then to arterioles and capillaries, the blood in the pulmonary system diverts into smaller and smaller vessels, ultimately ending at the capillaries. Once the blood enters these tiny vessels, it picks up the molecules of oxygen that are so important for cellular function. That process will be described in the section on the capillary system.

Venous System

On many levels, the venous system is the opposite of the arterial system. On the systemic side, the arterial system delivers blood away from the heart and to the tissues, and the venous system goes the other way, bringing the blood back to the heart. On the pulmonary side, the arterial system sends blood from the heart to the lungs for oxygenation, and the venous system sends the now-oxygen-rich blood back to the heart. Revisiting the road analogy, a traveler might leave a major city (the heart) via a superhighway (the aorta), then take a smaller highway (the arteries) and finally side roads (the arterioles) to get to a small town (the tissues) or other destination. To return to the city, the traveler would go in the opposite direction, beginning by taking the side roads, which are analogous to the venules; then the highways, or the veins; and finally the superhighway that leads into the city, or the heart. The venous system has two superhighways, which are two large veins called the superior **vena cava** and the inferior vena cava, which will be described later. (The plural of vena cava is venae cavae.)

Besides its role in returning blood to the heart, the venous system is a blood reservoir. Typically, about two-thirds of the body's circulating blood supply is in the venous system. The veins and venules temporarily store blood that can be immediately transported to the other areas of the body as necessary. Exercise, for example, initiates a series of responses within the body that affect the circulatory system. One occurs when **vasoconstrictor nerves** send messages primarily to the veins that tell them to constrict. As they do, blood moves from the venous system to the heart, which in turn sends added oxygenated blood to the arterial system and facilitates the increasing need for oxygen in the working muscles. Similarly, when a person loses a large volume of blood through a serious wound, nerves signal a reorganization of the blood from tissues that are less important to immediate survival to those tissues that are vital in maintaining life. The venous vessels constrict in some areas, such as the skin or the digestive tract. This forces blood to vessels in vital organs, such as the heart. The action not only preserves oxygenation to critical tissues, but also helps maintain the body's overall blood flow and helps ensure that the blood reaches its destination without delay.

Another important difference between the venous and arterial systems is the movement of the blood. Blood efficiently flows through veins and venules even though the driving **blood pressure** is lower than it is in the arterial system. Blood pressure is measured in a unit abbreviated as mmHg. It refers to millimeters (mm) of mercury (Hg), the standard method of measuring pressures. Scientists measure pressure by watching its effect on mercury—the silver liquid in old-fashioned thermometers. If mercury is placed in a tube, an increase in pressure will cause the mercury to expand and rise. A pressure decrease results in a drop of the mercury level. A reading of 120 mmHg means that the pressure in the aorta is enough to raise a column of mercury by 120 mm above the zero point, which is normal atmospheric pressure.

In the venous system, just 10–15 mmHg pressure is enough to force blood all the way from the venules back to the heart. This is about half of the average pressure of 30 mmHg seen even in the small arteries. The venous system can function on the lower pressure, which nears 0 mmHg (atmospheric pressure) by the time it gets to the heart, in part because the branching of the venous system is going in a direction opposite to that of the arterial system. Instead of an aorta that branches into smaller arteries and then miniscule arterioles—and slows as it goes—the venous "river" does the opposite. Blood from perhaps dozens of tiny venules merges into a slightly larger venule. When it does, the blood rate increases because the overall lumen (interior diameter) of the larger venule is still less than the combined lumens of all the smaller venules feeding into it. In other words, blood from a bigger space is squeezing into a tighter space. The same thing happens when venules merge into one small vein, or when numerous veins merge into the vena cava. The flow accelerates along the way much as a river's current hastens as new creeks join it.

The elasticity of the large and small venous vessels is also opposite to that of the arterial vessels. Whereas the largest arteries are the most elastic to tame the pulses of blood from the heart, the largest veins are the most rigid to help maintain or boost the blood flow. Conversely, the most distensible vessels on the venous side are the venules, which readily expand to serve as blood reservoirs.

The venous and arterial systems have other similarities and dissimilarities, too, and these will be discussed in the next sections.

Vessel Composition

Veins and venules have a three-layered structure much like that of arteries and arterioles (Figure 2.5). Each of these vessels has the same three layers: the tunica adventitia on the outside, the tunica media in the middle, and the tunica intima lining the inside. Just as in the arterial vessels, the muscular and elastic tunica media is the thickest layer in veins and venules and includes both elastin and muscle tissue to allow the vessel openings to expand and constrict. The elastin protein gives the vessels their stretch and slows the blood moving through them, while the smooth muscle cells contract and narrow the vessels' openings to urge the blood's pace.

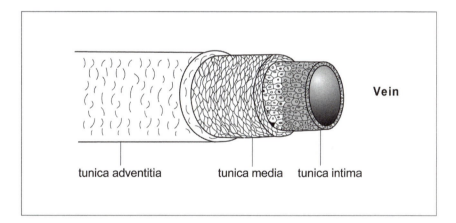

tunica adventitia   tunica media   tunica intima

Figure 2.5   Veins. Veins and venules have a three-layered structure: the tunica adventitia on the outside, the tunica media in the middle, and the tunica intima lining the inside. The main difference between these vessels and those of the arterial system is in the middle layer. The veins and venules have a thinner tunica media with a higher collagen content and with fewer muscle cells that are commonly in a less-ordered arrangement. (Sandy Windelspecht/Ricochet Productions)

Collagen in this middle layer helps to rein in the vessels' elasticity and yields structural strength.

The tunica adventitia, or outer layer, of veins and venules is a covering of connective tissue. Its job is to hold the vessel in place. The innermost of the three layers, or the tunica intima, is a one-cell-thick sheet of endothelial tissue.

So far, this description of the composition of veins and venules is the same as that for arteries and arterioles, but differences do exist. In the arterial system, the delineations between the three layers are much more distinct than they are in the venous system. The transition in venules and veins is rather gradual. In the arteries and arterioles, the smooth muscle cells of the tunica media wrap around the vessels in a very regularly arranged fashion. The veins and venules have a thinner middle layer with fewer muscle cells that are commonly in a less-ordered arrangement. This would result in a severe lessening of their mechanical strength were it not for their high collagen content. The additional collagen typically seen in veins and venules also limits the vessels' elasticity somewhat. As seen in the arterial system section, arterial vessels, especially the larger arteries, must distend to even out the strong blood pulses from the heartbeat and regulate the blood flow. There is no similar demand on veins and venules, but they do need to stretch out enough to accommodate the temporarily stored blood they are holding. In effect, veins and venules are striking a balance between stretching to hold up to 70 percent of the body's blood and remaining inelastic enough so that they can maintain a sufficient flow to transport the blood back to the heart.

Venous blood faces other challenges on its way back to the heart. The force of gravity discourages blood flow. In a standing individual, the blood in the feet has to overcome the gravitational pull to flow up the leg and back to the heart, and it does it without the heart's pumping action to help it along. While fighting gravity, the blood also must flow into increasingly bigger vessels, which would seemingly have the effect of slowing blood flow. In large part, blood can overcome these obstacles for the same reason that venules develop a flow with the influx of capillaries: Blood is still moving from a larger space to a smaller space. Because so many smaller vessels feed a larger vessel, the combined cross-sectional area of the smaller vessels is greater than that of the larger vessel, and blood flow actually speeds up as it approaches the heart. In addition, venous vessels become stiffer as they get closer to the heart, because their collagen content is greater. This ensures that the vessels will not distend and therefore will not create a larger space for blood or slow the blood rate.

Systemic Circulation

The systemic side of the venous system begins at the capillaries and ends at the heart. The capillaries are the site of blood-to-tissue transfer of nutrients and other materials, including oxygen. Once that oxygen and nutrient transfer has occurred at the tissues, the venous system takes over to collect the blood from the capillaries and convey it back to the heart. Oxygenated blood in the arterial system is bright red, because oxygen causes a change in the three-dimensional configuration of the large compound called hemoglobin that is found in red blood cells. Without the oxygen, the blood appears dark maroon. For this reason, blood leaving the capillaries—after the oxygen drop-off—is more blue than red. The venules are the first vessels in the return of blood from the capillaries to the heart. A typical venule has a lumen of about 7.9×10^{-4} inches (20 μm) with a vessel wall that is about 7.9×10^{-5} inches (2 μm) thick. This compares to the average arteriole, which has a like-sized lumen but a wall thickness of 5.9×10^{-4} inches (15 μm). Many capillaries, which are about a quarter of the size of a venule, may empty into a venule, and, as already described, this onslaught helps to generate an increased flow. From there, smaller venules merge into larger and larger venules, and eventually into small veins. Although the size varies considerably, a typical vein has a lumen of around 0.197 inches (5 mm or 5,000 μm). Its wall thickness averages about 0.02 inches (0.5 mm or 500 μm). When looking at the circulatory system as a whole, the venous vessels often have arterial counterparts, with blood flowing out to the tissue in an artery and back to the heart in a nearby, sometimes adjacent, vein.

This chapter will now provide a closer view of some of the major veins in different body areas, beginning with the leg. Blood from the foot may ascend into the anterior tibial vein, which is named for the tibia (one of the lower leg bones). Veins in the ankle and numerous capillaries in the leg empty first into peroneal veins, which are also known as fibular veins because of their location in the region of the fibula (the other lower leg bone), and then into the posterior tibial vein, which eventually unites with the anterior tibial vein. Both the anterior and posterior tibial veins, which also accept blood from numerous other capillaries in the lower leg, flow into the popliteal vein at the back of the knee. The popliteal vein, in turn, empties into the femoral vein, a large vessel in the thigh (alongside the large upper leg bone, or femur). Blood from the foot can also take a more direct route to the femoral vein by way of the great saphenous vein, the longest vein in the human body.

Regardless of how it reaches the femoral vein, all of this blood flows into the external iliac vein. Now in the abdomen, blood from the external iliac vein joins with the internal iliac vein, which carries blood from the pelvis to form the common iliac vein. This vein joins the large inferior vena cava, one of the two "superhighways" in our road system bringing blood back to the heart.

Many other major veins in the abdominal cavity empty directly or indirectly into the inferior vena cava. Within the blood supply for the reproductive system, for example, the female body has a pair of ovarian veins, and the male body has a pair of spermatic veins. Both the right ovarian vein and the right spermatic vein connect directly to the inferior vena cava, but the left ovarian and spermatic veins first merge with the left renal (kidney) vein, which then unites with the inferior vena cava. The female's uterine veins take a more convoluted route to the inferior vena cava, first merging with the internal iliac vein that unites with the common iliac vein, and finally joining the inferior vena cava.

Other major body areas that use the inferior vena cava include the kidneys, liver, spleen, digestive system, and pancreas. The renal veins of the kidneys and the hepatic veins of the liver empty directly into the inferior vena cava. The liver performs a vital function because it absorbs products that the blood has gained from the digestive system. For example, the liver absorbs glucose (a sugar that results mainly from starch digestion) and uses much of it to make glycogen, which is basically a storable form of glucose. When the body needs extra energy, the glycogen transforms back into glucose.

Blood from the spleen takes a less direct route to the inferior vena cava. It drains from the spleen via the splenic vein, which joins the superior mesenteric vein to create the portal vein. The portal vein empties into the liver. Venous blood leaves the liver through the hepatic veins, as previously described, and discharges into the inferior vena cava.

The digestive system has many veins emptying the intestines, rectum, stomach, and other specific areas. These veins include the rectal, pudendal, lumbar, superior and inferior mesenteric, gastric, gastroepiploic, and epigastric veins. Each has its own path to the inferior vena cava. As an example, the rectal veins number three: the inferior, middle, and superior rectal veins. The inferior rectal vein joins the internal pudendal vein, which flows into the internal iliac vein, while the middle rectal vein connects directly to the internal iliac vein. The internal iliac vein then continues to the common iliac vein, which finally unites with the inferior vena cava. The superior rectal vein avoids the internal iliac vein altogether and instead drains into the inferior mesenteric vein, which flows into the splenic vein, then to the superior mesenteric vein and portal vein, through the liver, into the hepatic vein, and to the inferior vena cava. As the inferior vena cava returns blood from the lower body to the heart, another superhighway is doing the same for the upper body. This large vein is the superior vena cava, which gathers blood from the arm, chest, head, and neck regions.

The arm's arrangement is somewhat similar to that of the leg. The veins, of course, are named differently to reflect the specific body area in which they are found. Veins in the hand flow into the radial vein, the ulnar vein, or the basilic vein. The radial vein eventually merges into the ulnar vein, which then continues into the brachial vein in the upper arm. The basilic and brachial veins flow into the axillary vein that carries blood into the chest. Just as some blood from the foot can patch nearly directly into the femoral vein via the long saphenous vessel, some of the blood from the hand can drain through a long cephalic vein right to the axillary vein of the upper arm. The axillary vein flows into the subclavian vein and then the brachiocephalic vein (also called the innominate vein) of the upper chest.

The head and neck region have several major veins, but the most well known are the jugular veins, which accept blood from the brain, face, and neck. These veins flow into either the subclavian or brachiocephalic veins. One of the three jugular veins, the internal jugular vein, is the largest venous vessel in the head and neck. This vessel actually merges with the subclavian vein to form the brachiocephalic vein of the upper chest. Blood from tissues of the chest muscles, from the thyroid gland, and from the diaphragm also either directly or indirectly release into the brachiocephalic vein. The brachiocephalic vein drains into the superior vena cava. The ultimate destination of both superhighways—the inferior vena cava and the superior vena cava—is the heart. Specifically, they deliver blood to the right atrium of the heart.

Pulmonary Circulation

Like the arterial system, the pulmonary side of the venous system is much simpler than the systemic side. Just two major veins are involved: the bronchial vein and the pulmonary vein. Most veins are single or paired, but the pulmonary veins are actually four vessels that flow directly from the lungs to the left atrium. These veins return newly oxygenated blood from the lungs to the heart. The bronchial veins, on the other hand, drain blood of the bronchi and a portion of the lungs, then travel through one or more smaller veins to reach the superior vena cava, which brings blood into the heart's right atrium.

Capillary System

Although the **capillaries** are the smallest vessels in the circulatory system, they represent the main exchange site between the blood and the tissues (Figure 2.6). They can be viewed as both the ultimate destination of the arterial system and the starting point of the venous system. From the heart, blood travels through the arteries to the arterioles, and then to the capillaries, where exchange occurs. Nutrients, oxygen, and other materials carried by the blood are traded for waste products from tissue cells. Blood continues down the capillaries, soon entering the venules and then the veins on its return trip to the heart.

Vessel Composition

Capillaries have a composition that is somewhat different from that of other circulatory vessels. As already described, arteries, arterioles, veins, and venules have a three-layered construction, including elastic tissue

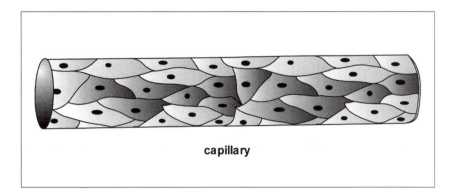

capillary

Figure 2.6 Capillaries. Capillaries have an important function in exchanging gases and materials between blood and tissues, but their composition is quite simple. A capillary is basically a tube comprised of a single layer of epithelial cells. (Sandy Windelspecht/Ricochet Productions)

and smooth muscle. Capillaries, in contrast, are composed of just a single layer of endothelial cells, which gives them a wall thickness of just 2.0×10^{-5} to 3.9×10^{-5} inches (0.5–1 µm). Their internal diameter, or lumen, is about 1.6×10^{-4} to 3.9×10^{-4} inches (4–10 µm). That is a tight fit for red blood cells, which range from 2.6×10^{-4} to 3.5×10^{-4} inches (6.5–8.8 µm) in diameter. To squeeze their way through these tiny vessels, red blood cells travel in single file (a so-called bolus pattern), and when that is not enough, they bend, twist, partially fold, and otherwise deform. At the same time, the capillaries distend to allow the blood cells to pass through them. It is a tortuous pathway, but it is a short one: Capillaries are typically just 0.02–0.039 inches (about 0.5–1 mm) long.

Capillaries come in three types:

- Continuous

- Fenestrated

- Discontinuous or sinusoidal

Continuous capillaries are constructed of epithelial cells that overlap tightly, leaving no gaps between them. They are present in the skin, muscles, and lungs as well as the central nervous system (the brain and spinal cord), and are the least permeable type of capillary. Only those substances with molecular weights of less than 10,000 can easily cross them. They are particularly important in what is known as the blood-brain barrier. This barrier prevents damaging substances from being transmitted from the circulating blood to brain tissue and to the watery cerebrospinal fluid that cushions and protects the brain and spinal cord. Although continuous capillaries generally permit only small molecules to traverse them, even large molecules with molecular weights of up to 70,000 can make their way across, given enough time. Scientists believe this is accomplished through temporary openings that may occasionally form between epithelial cells.

Unlike continuous capillaries, fenestrated capillaries are always full of holes. The endothelial cells overlap much less tightly, creating gaps. In addition, they have numerous pores, or fenestra (literally, "little windows"), of 2.0×10^{-6} to 3.9×10^{-6} inches (50–100 nm) in diameter. These openings greatly increase the capillaries' permeability, making them particularly useful in tissues that exchange a great deal of fluid and metabolites with the blood. Fenestrated capillaries are common in such tissues as the kidney and the **intestinal villi**, which are tiny projections that serve as the nutrient exchange point for the intestines.

Discontinuous or sinusoidal capillaries are large capillaries with apertures so wide that bulky proteins and even red blood cells can pass through them. Of the three types of capillaries, they are the most permeable to water and solutes. They are found in such tissues as the liver, spleen, and bone marrow.

Exchange Function

The exchange of water, oxygen, nutrients, hormones, drugs, waste products, and other chemicals occurs primarily at the capillaries. Oxygen and carbon dioxide, which are gases, move by passive molecular flow, a process called **diffusion**, right through the wall. An individual endothelial cell, like other cells in the human body, is surrounded by a thin membrane made of two fatty layers, the lipid bilayer. Substances that can pass through this layer are termed lipophilic (fat-soluble). Besides oxygen and carbon dioxide, other substances that can readily cross the membrane include some drugs, like the general anesthetic a person receives before surgery. Actually, oxygen and other materials do not move directly from capillary to cell or vice versa. Instead, they first enter a region just outside the cell. This fluid-filled extracellular area is called the **interstitial space**.

The direction of diffusion is determined by the **concentration gradient**, which means that molecules travel from an area of high concentration to one of low concentration. Transport from high to low concentration is described as moving down the concentration gradient. Therefore, if an arteriole delivers oxygen-laden blood to a capillary near oxygen-poor tissue, the oxygen (O_2) will pass from the blood to the tissue, or from an area of high concentration to an area of low concentration. The same thing happens with the waste product carbon dioxide (CO_2), which accumulates in tissue. The carbon dioxide moves from the cell to the blood, or from an area of high concentration to an area of low concentration. The reverse likewise occurs in the lungs, when oxygen-poor blood arrives to pick up oxygen from and drop off carbon dioxide to the alveoli, the tiny air sacs in the lungs. In this case, the alveoli gather oxygen with every breath a person takes, then deliver it to the blood in the pulmonary circulation. Oxygen in the blood is carried by the large hemoglobin molecule, also known as a respiratory pigment. Each hemoglobin molecule can carry four oxygen molecules. Because red blood cells in the average adult number from 24.8 to 28.6 trillion, and each red blood cell contains about 300 million molecules of hemoglobin, the potential for oxygen transport by the blood is immense. (See Chapter 10 on the respiratory system for more information about this process.)

The oxygen exchange between the capillaries and the alveoli, as well as other tissue cells, is possible because of their proximity to one another. The time it takes for a molecule to move is proportional to the square of the distance moved. The time quickly escalates as oxygen has farther to go. For this arrangement to work in the human body, an enormous number of capillaries is required—enough to place capillaries within 3.9×10^{-4} to 7.9×10^{-4} inches (10–20 m) of just about every cell—just one, two, or three cells' diameter away. Some organs even have special networks or clusters of capillaries. In the kidney, the cluster is known as the glomerulus and facilitates the considerable volume of blood that flows to and from this organ. The capillaries in the glomerulus are typically fenestrated.

(See Chapter 12 on the urinary system for more details.) Other tissues and organs that have a higher density of capillaries include the heart and skeletal muscles, which are both very metabolically active and demand a highly proficient transfer of gases and nutrients. In organs and tissues, such as joint cartilage, that are less active and less oxygen-demanding, fewer capillaries are necessary.

Of course, the body cannot rely solely on diffusion to get oxygen all the way from the lungs to every tissue cell. The pumping of the heart does much of the transportation, forcing the blood along the arteries, then more slowly into the arterioles, and, slower yet, into the capillaries. Blood is progressing so slowly in the capillaries that each red blood cell commonly takes 1–2 seconds to traverse it, ample time for diffusion to occur.

The Blood Circuit

Now that the basic components and functions of the arterial, venous, and capillary systems have been described, this section will take a closer look at how the blood makes the transit through those vessels.

One-Way Valves

Blood valves are located throughout the circulatory system. Just as a door marked "push" or "pull" only opens in one direction, these valves swing strictly one way. In the arterial system, blood rushes from the left side of the heart through a valve and into the aorta. As the heart beats, the valve swings wide to let the blood pass into the aorta. The blood flow naturally slows when the contraction ends, and the valve swings shut. This prevents the blood from streaming back into the heart chamber.

Other large blood vessels have similar valves that open in just one direction. In the legs, for example, a pair of semilunar valves swing outward to the vessel walls, allowing blood to move past. Any reverse flow causes the valves to close. Just as wind from outdoors may cause a door to swing open, but a breeze from indoors can quickly slam it shut, the valves allow blood to move one way, but not the other. This simple system ensures that deoxygenated and waste-filled venous blood doesn't flow backward and mix with the oxygenated, nutrient-filled arterial blood.

Blood Pressure

Systemic Circulation

Blood pressure is a key force driving the blood through the arterial system. Blood leaves the heart in the systemic circulation under a very high pressure caused by the heart's contraction. On average, the blood pressure in the aorta reaches 120 mmHg following a heartbeat, then falls back down to 80 mmHg before the next heartbeat. This is often written as 120/80 mmHg.

The 120 mmHg reading in the aorta immediately following a heartbeat is the highest pressure that blood reaches in the circulatory system. As the heart's contraction ends, the blood pressure quickly drops. As seen earlier, the aorta and other large arteries have a high percentage of elastin, the protein that permits stretching. When the pulse of blood enters the aorta, that large vessel quickly distends, then slowly returns to its normal size. By doing so, it eases the blood pressure. Subsequent arteries do the same thing, although they become less and less elastic as blood moves farther from the heart. Every time a vessel widens, the pressure drops a bit. By the time the blood reaches the junction between the small arteries and arterioles, it has diminished to about 60–70 mmHg. At the arteriole-capillary border, the blood pressure is just 35 mmHg.

Pulmonary Circulation

The pulmonary circulation begins with the beat of the right side of the heart, which is smaller and less powerful than the left side that drives the systemic circulation. Here, the pressure of the blood leaving the heart and entering the pulmonary artery is just 25 mmHg, which is still sufficient to force the blood the small distance from the heart to the lungs. Between heartbeats, the pressure drops to about 10 mmHg. As in the systemic circulation, the pressure continues to decline as the blood travels from arteries to arterioles to capillaries.

The Return Trip

If the arterial system requires the force of the heartbeat to drive blood to the lungs and body tissues, what does the venous system use to propel the blood back to the heart? The answer has several parts.

As shown in the section on the venous system, the veins and venules greatly outnumber the arteries and arterioles. They are also oriented in the opposite direction with blood flowing from the smallest and most plentiful vessels into ever-larger but fewer vessels. In the arterial system, the total cross-section of the vessels (the sum of their lumens) increases as they get farther from the heart, which results in a slower blood velocity. This reliance of the flow rate on cross-sectional area is illustrated in the mathematical equation:

$$\text{velocity of the blood (mm/sec)} = \text{blood flow (mm}^3\text{/second)/} \text{cross-sectional area (mm}^2)$$

The velocity is the overall speed of the blood through a vessel, the blood flow is the volume of blood that moves through a vessel, and the cross-sectional area is the size of the vessel's lumen. The mathematical formula shows that velocity is inversely proportional to the cross-sectional area, so as cross-sectional area increases, the velocity decreases.

In the venous system, the vessels increase in size closer to the heart, but their number decreases dramatically, until the two large venae cavae are accepting blood from thousands of vessels throughout the entire venous system. The sum total of cross-sectional areas of smaller vessels greatly surpasses the area of larger vessels, which results in an increase in blood velocity. In summary, both the vessels of the arterial and venous systems are largest and least numerous near the heart, and smaller and more numerous away from the heart. The difference is in the direction of flow. On the arterial side, blood is moving away from the heart and slows as it goes. On the venous side, blood is returning to the heart and speeds up as it approaches the heart.

In addition, venous blood gets a little help in its return trip from the structure of the vessels, from muscles, and even from arteries. By the time blood makes its way through the arterial system and the capillaries, its pressure in the venules is just 15 mmHg and is nearly nonexistent by the time it reaches the venae cavae. With such little impetus to return to the heart, gravity would influence the blood to pool wherever the body is closest to the ground. This usually does not present a debilitating problem because venous vessels have the ability to deflate and reinflate with very little pressure applied to them, so even the small 15 mmHg gradient from venules to heart is sufficient to urge the blood along. The arteries also help. Because arteries and veins are usually in close proximity, the strong pulses of blood that move from the heart and down the arteries put some pressure on the adjacent veins and help circulate the blood. In addition, the venous blood near the heart gets an added incentive from the heart itself. As the heart's valves open to allow in venous blood, suction results and actually draws the approaching blood into the waiting chamber.

The body's skeletal muscles also help by contracting and relaxing, and pressing on venous vessels. This action squeezes the blood back to the heart through the re-inflated veins. This type of muscle action produces obvious results in an exercising person, but even when a person is quietly standing, these muscles are continually contracting and relaxing, and pushing blood into veins. Valves in the larger veins assist as well. Once the blood makes its way partially up the leg, the valves prevent it from rushing back down. In the large leg veins, valves occur about every half inch (1.25 cm) along the vessel.

The legs are not the only parts of the body that have to contend with the effects of gravity. When a person stands up, about 16.9 ounces (500 ml) of blood shifts to the lower legs, so some pooling does occur. The volume of liquid lowers as plasma from the blood filters out of the vessels and into adjacent tissues. This causes an overall drop in blood volume, and the body responds by lowering cardiac output. As a result, less blood reaches other parts of the body, including the brain. Usually the body responds by quickening the heart rate, constricting the venous vessels in the legs, or using nervous control to decrease the amount of blood reaching the legs. When these steps are not enough, a person begins to feel light-headed. By putting the head between the knees—or, in extreme examples, by fainting—the head is lowered, making it much easier for the blood to fight gravity and reach the head.

Effect of Velocity on Capillaries

The blood's flow rate is also an important consideration in the blood-to-tissue exchange that occurs in the capillaries. Because of capillaries' distance from the heart and the distance of the arterial vessels, as well as the large cross-sectional area of the capillaries, the blood's velocity is so slow that blood cells barely creep through the capillaries. In fact, the blood's velocity in the capillaries is less than 1/200th its speed in the aorta. The sluggishness provides ample time and optimal conditions for diffusion of oxygen and other materials to occur. In most cases, blood cells take 1–2 seconds to traverse a capillary, a considerable time for a vessel that is just 0.02–0.039 inches (about 0.5–1 mm) long.

Control of Blood Flow

Although the circulatory system is similar to a road system in some regards with the vessels analogous to the highways and streets, the cardiovascular system is hardly passive. The blood vessels are alive and dynamic structures that can change diameter and alter the flow of the blood within them.

As previously shown, the blood vessels contain smooth muscle. When contracted, these muscle cells can narrow blood vessels. When the contraction ends, the blood vessel returns to its larger diameter, which is driven by the pressure of the passing blood. Even when the smooth muscle cells are not contracting, however, they are imparting muscle tone to the vessels.

The mechanism for the control of this tone is called the **Bayliss myogenic response**. Without this response to counteract the force of the blood pressure, the vessels would continue to stretch wider and wider, which would affect flow.

Besides those in the peripheral venous system, other blood vessels change diameter to meet the needs of the body. These types of adjustments occur continually. For example, the brain requires a constant, sufficient blood flow. Oversight and maintenance of this flow is the job of the **arterial baroreceptor reflex**, which responds to slight changes in blood pressure. In this case, the **baroreceptors**, which are pressure detectors located in the major arteries, sense a dip or spike in blood pressure. When pressure increases, the baroreceptors reflexively send a message that shuts down the vasoconstrictor center in the brain's medulla, which is responsible for constricting blood vessels. With the vasoconstrictor center offline, the blood vessels dilate and pressure drops. At the same time, cardiac output changes to meet demands and venous vessels likewise dilate. These actions collectively return blood pressure to normal. If, however, the blood pressure remains elevated, as occurs in a person with **hypertension** (high blood pressure), the baroreceptors become accustomed to the new, higher average pressure and stop sending messages to the vasoconstrictor center. As a result, the body no longer tries to rectify the higher blood pressure.

The arterioles play a vital role in controlling blood flow. These vessels are often called resistance vessels because they offer resistance to and therefore regulate the flow of blood from the heart. Due to the considerable smooth muscle in the walls of these vessels, the lumen can change diameter to allow more or less blood to reach the tissues. In addition, the arterioles that feed capillaries, called **terminal arterioles**, take their cues from local metabolic factors rather than nerves. Each terminal arteriole serves as a door to its own little network of capillaries. Based on metabolic needs, the arteriole can close tight or open wide to regulate blood flow.

Venous muscle tone is also important in peripheral venous vessels, which regulate blood volume by adjusting the amount of liquid in the plasma and in such tissues as the skin, muscles, and kidneys. These peripheral vessels can thus act as blood reservoirs that can temporarily hold unneeded blood, which has a regulating effect on the overall blood supply.

Hormones regulate blood flow, too. Adrenaline is perhaps the most well-known hormone in this regard. When a person is under physical or mental stress, the body boosts adrenaline production. This hormone, as well as angiotensin II and vasopressin, helps to maintain a healthy blood pressure even under the most dangerous conditions, like severe blood loss. (Hormones are discussed further in the endocrine system chapter in this encyclopedia.)

In addition to controlling blood flow through muscles, hormones, and nerves, the body has fail-safes in certain organs, including the brain. Here, arteries and arterioles merge to create what are called **arterial anastomoses**. These are alternative sites where the organ can obtain blood if a supplying artery becomes blocked. The anastomoses essentially serve as backup blood sources for use in emergencies. (The arterial anastomoses are not to be confused with arteriovenous anastomoses, which are wide vessels found especially in the skin and nasal mucosa. These vessels bypass capillaries and attach arterioles and venules directly.)

The Heart: A Living Pump

Although it is only as big as a fist, the heart drives the circulatory system. This organ works constantly over the course of a lifetime, pumping blood to the lungs in the pulmonary circulation and to all other body tissues in the systemic circulation. Even a short pause in its functioning can result in death. This section will take a closer look at this amazing organ and how it carries out its job.

Anatomy and Blood Flow

The heart is truly a pump made of muscle tissue. Blood moves in and out of this pump through four chambers inside (see Figure 2.7). The two, smaller, top chambers are called atria (atrium is the singular form) and two bottom

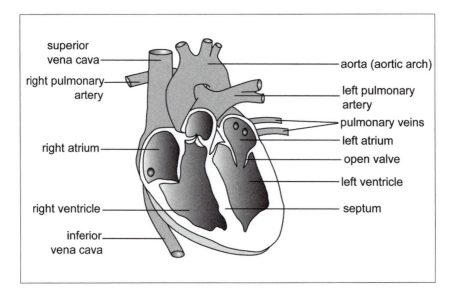

Figure 2.7   The Heart. The heart has two upper chambers, called the atria, and two lower chambers, or ventricles. The right and left sides of the heart are separated by a thick wall, known as the septum, and the atrium and ventricle on each side are divided by valves that open to allow the passage of blood. The superior and inferior venae cavae are the main blood vessels that return deoxygenated blood from the upper and lower body, respectively, to the heart's right atrium. The right and left pulmonary arteries deliver blood from the right ventricle to the lungs for oxygenation, and the newly oxygenated blood returns to the left atrium through pulmonary veins. Newly oxygenated blood leaves the heart by way of the aorta, the largest artery in the body. Near the heart's right ventricle, the aorta bends. This is known as the aortic arch. (Sandy Windelspecht/ Ricochet Productions)

chambers are called ventricles. It actually works like a double pump, with the right half taking in blood returning to the heart from the body tissues, and then sending it over to the lungs to drop off carbon dioxide and pick up oxygen. The newly oxygenated blood then heads back to the other side of the heart, where it gets the added boost it needs to propel it to the body tissues again. This double pump springs into action an average of 70–80 times every minute, all day and all night, for a person's entire lifespan.

Cardiac Muscle

The heart is made out of cardiac muscle (also known as myocardium), a tissue that is unlike the smooth or striated muscle seen elsewhere in the human body. Striated muscle is the tissue that a person uses to move his or her legs or fingers. Because the individual can control it, it is also known as **voluntary muscle**.

This tissue has light and dark bands, called striations, which give skeletal muscle yet another name: striated muscle. Smooth muscle, like that in blood vessels, is known as **involuntary muscle**, because a person cannot direct its movements like he or she can control skeletal muscle. Instead, the autonomic nervous system controls its action. Falling somewhere in the middle of these two types of tissue is cardiac muscle. Cardiac muscle has the striations seen in skeletal muscle, but it takes its direction from the autonomic nervous system like the smooth muscle does. Unlike either striated (also known as skeletal) or smooth muscle, cardiac muscle cells are very closely linked to one another and have fibers that interconnect one cell to the next. As will be shown in the section on electrical activity later in this chapter, this is vital in making the heart beat as a unit. In addition, cardiac muscle does not tire out like skeletal muscle does, and it requires a shorter resting time between contractions. It is easy to assume that skeletal muscle can contract for a very long time, especially when considering how a body maintains muscle tone. A closer look reveals that different groups of skeletal muscle alternately shorten to give the appearance of constant contraction, even when the muscle cells are individually contracting and relaxing. In the heart, conversely, all of the cardiac cells contract at the same time. (For an in-depth discussion of cardiac muscle, see Chapter 11 on the musculoskeletal system.)

This muscle tissue of the heart surrounds all four of its chambers, completely enclosing them. The muscular walls of the atria do not require the kind of force that the ventricles do, and they are considerably thinner. All of the chambers are lined with endocardium, a thin membrane that provides a smooth, slick surface for the blood to slide along. This membrane is similar to the endothelium that coats the inside of blood vessels. The entire heart is enclosed in a fluid-filled fibrous sac, called the **pericardium**, that attaches to the diaphragm. Actually a double sac with fluid between the two layers, the pericardium supports, lubricates, and cushions the heart. The diaphragm, which is a large muscle that separates the chest and abdominal regions, pulls downward during inhalation and consequently tugs the heart into a more upright position.

As mentioned, the atria sit at the top of the heart and the ventricles at the bottom. The word atrium means "entrance hall" in Latin. These chambers got their name because the blood enters the heart through the atria. At one time, atria were known as auricles because they somewhat resemble ears, and auris is Latin for ear. Ventricle is Latin for "little belly," and one might say they look like ministomachs. If the atria are considered the entrances to the heart, the ventricles are the exits, and blood drains from them via the pulmonary artery and aorta. The right and left sides of the heart are completely separated by a thick muscular wall called the septum. This ensures that deoxygenated and oxygenated blood do not mix.

In addition, the heart has a ring of fibrous connective tissue, called the **annulus fibrosus**, that serves as an anchor for the heart muscle and as an almost continuous electrical barrier between the atria and ventricles. Electrical charge plays a key part in proper heart functioning and will be discussed later in this chapter.

The right atrium is separated from the right ventricle by a valve. The same holds true for the left atrium and left ventricle. This arrangement allows blood to move from one to the other, but only when the valve is open. Like the valves in the blood vessels, the heart valves permit blood flow in only one direction. Between the two right chambers, the valve has three flaps, or cusps, and is known as a tricuspid atrioventricular valve. It is commonly called the AV valve. A bicuspid (two-cusped) mitral valve (also known as the mitral valve) divides the two left chambers. The ventricles have additional valves to seal off flow between them and the arteries. The valve between the right ventricle and pulmonary artery is called the pulmonary semilunar valve, and the valve between the left ventricle and aorta is called the aortic semilunar valve. Both have three cusps.

The AV, mitral, and semilunar valves all work the same way. When blood is rushing through a valve, each of which is about 0.004 inches (0.1 mm) thick and made of fibrous connective tissue lined with membranous tissue, its cusps open with the flow. As the blood tapers off, the cusps fall back to their original closed positions. Tiny tendinous cords, appropriately called **chordae tendineae**, attach to adjacent muscles (papillary muscles) and prevent the valve's cusps from falling back too far and letting blood seep through.

Blood Flow and the Heart

The valves, chambers, and muscle tissue are all involved in blood flow through the heart. Venous blood returning from the body tissues enters the heart through the two large venae cavae. The anterior vena cava carries blood from the head, neck, and arms, and the posterior vena cava conveys the blood from the rest of the body tissues (except the lungs). The right atrium fills with blood from the venae cavae, as well as the coronary sinus, which is the main vein carrying blood from the heart (details of the coronary circulation are included in the next section). When it contracts, the pressure of the blood builds, forcing open the tricuspid AV valve and allowing the blood to flood the right ventricle. When the atrium relaxes, the blood's pressure drops, and the AV valve falls back to its original position. The now-filled ventricle contracts. The only way out for the blood is through the pulmonary semilunar valve, which opens outward into the pulmonary artery.

Blood rushes out of the heart and into the pulmonary artery, where it branches and eventually reaches the lungs and alveoli to pick up oxygen. Following the ventricle's contraction, the blood pressure in the pulmonary artery is 35 mmHg. The newly oxygenated blood returns to the left side of the heart and enters the atrium. It contracts, forcing blood through the bicuspid mitral valve in the left ventricle. Now that the blood is in the left ventricle, the mitral valve closes, the ventricle contracts, and blood flows through the aortic semilunar valve and into the aorta. The force of the contraction boosts the blood pressure in the aorta to 120 mmHg, more than three times the pressure of a similar volume of blood in the pulmonary artery. To accomplish this feat, the muscular wall of the left ventricle is about three to four times as thick as the right ventricle's wall. The increased thickness gives the left ventricle the power it needs to drive blood throughout the body. Normally, the amount of blood in a completely filled adult ventricle is about 0.12 quarts (120 ml). This amount is called the **end-diastolic volume**. The heart typically only ejects about two-thirds of this blood, retaining 0.04–0.05 quarts (40–50 ml). That yields a total stroke volume—the amount of blood that exits the heart—of 0.07–0.08 quarts (70–80 ml). When the body needs to increase stroke volume, such as during periods of heavy exercise, it calls on the heart to begin pumping the residual 0.034 quarts (40–50 ml).

As it turns out, both atria fill simultaneously, so the right side of the heart is taking in deoxygenated blood from the systemic circulation at the same time that the left side is admitting newly oxygenated blood from the pulmonary circulation. The ventricles likewise fill at the same time. When either the ventricles or the atria contract, the chambers get smaller. When the atria relax following a contraction, they go back to their normal, larger size. The ventricles have an effect on the size of the atria, because they contract when the atria are resting. As the ventricles contract or shorten, they actually pull down on the bottom of the atria to further expand these upper chambers. This enlargement creates suction in the atria and serves to draw in blood from the venae cavae and begin preparing the way for the next heartbeat. Cigarette smoking can have a damaging effect on the circulatory system, as detailed in Sidebar 2.2.

The heart spends about as much time resting as it does contracting, with each contractile and resting period lasting less than half of a second. The contractile period is known as **systole**, and the resting period is known as **diastole**. Although each contraction lasts only about 0.4 seconds, that is enough time for the atria and the ventricles to contract. Careful observation reveals that the atria contract slightly before the ventricles during systole, so that blood travels from the atrium to the ventricle and to the aorta or pulmonary artery with every contraction.

The heartbeat's familiar "lubb-dupp" sound is actually the valves vibrating when they flap shut rather than the heart muscles expanding

SIDEBAR 2.2

The Damaging Impact That Smoking Has on the Circulatory System

It is estimated that cigarette smoking leads to one out of every five deaths every year in the United States. It damages nearly every organ of the body, particularly the organs and components of the circulatory system, such as the heart and blood vessels.

Cigarettes contain tobacco, and there are chemicals in tobacco that shrink the blood vessels, leading to a condition know as atherosclerosis, which can lead to coronary artery disease (CAD). CAD occurs when plaque builds up in the blood vessels, making it difficult for blood to get to the heart. Smoking is also a risk factor for peripheral arterial disease (PAD), which occurs when plaque builds in the vessels and prevents blood from traveling to the head, limbs, and other organs in the body.

It is not only heavy smokers who are doing damage to their bodies. Research has shown that even occasional smoking can cause significant damage to the circulatory system, as well as the rest of the body. Female smokers who take oral contraceptives are at greater risk for developing CAD, as are patients who have certain chronic conditions such as diabetes. Secondhand smoke—inhaling smoke from nearby cigarettes, cigars, or pipes—can also be harmful to the heart and blood vessels. In fact, it is estimated that secondhand smoke can lead to approximately 50,000 deaths every year in the United States. Most of these deaths are due to heart disease.

How to reduce this risk? Avoid tobacco smoke. Because tobacco has addictive properties, medical experts recommend that people never start smoking or quit immediately. In addition, secondhand smoke should be avoided, given the damage that this exposure can do to the body. While quitting smoking can be hard, there are numerous strategies, programs, and medicines that can help with the quitting process.

and contracting as many people think. These vibrations (about 100 cycles per second, or 100 hertz) result when the heart's contraction ends and the blood starts to slosh back toward the valves. As the liquid hits the valves, they shudder slightly. That shudder is the vibration that is audible to a doctor with a stethoscope pressed to a patient's chest. The "lubb" half of the heartbeat is the closing of the tricuspid AV and mitral valves that separate the atria and ventricles. The "dupp" is the sound made by the semilunar valves located between the ventricles and either the aorta or pulmonary artery.

Coronary Circulation

The heart is a hardworking muscle that demands its own arteries and veins to maintain its operation. In fact, some textbooks and medical articles even describe the human body as having three circulatory systems: the systemic, the pulmonary, and the **coronary circulations**.

The two major arteries feeding the entire heart muscle are the right and left coronary arteries that stem from the base of the aorta. The right coronary artery primarily delivers oxygen-rich blood to the two right chambers: the right atrium and the right ventricle. The left coronary artery mostly feeds the left side. Unlike the right artery that remains as a single, large vessel, the left almost immediately splits into two vessels, known as the transverse and descending branches. Each artery also ships a little blood to the opposite side, but the right artery mainly concentrates on the right side, and the left artery on the left side. Strangely, the amount of blood delivered by the two sides differs in individuals. About half of all people have a dominant right artery, 3 in 10 have equal delivery in the two arteries, and about one out of five have a dominant left artery.

As mentioned, the heart demands a strong blood flow to supply the oxygen it requires. The body will even respond to a blockage in a coronary artery by rerouting blood through nearby **collateral arteries** and around the compromised area. Heart patients frequently refer to this phenomenon as "growing new arteries." In addition, the heart has more than 2,000 capillaries per 0.00006 inch (mm3) that help ensure an ample oxygen supply to this active muscle.

Electrical Activity

How does the heart keep pumping? What triggers a heart to beat? Electrical activity is the answer. In the human body, the nervous system controls the overall electrical activity, including that in the heart, and is thus its primary regulator.

In skeletal muscle, nerves outside the muscle direct its contraction. In the heart, small and weakly contractile modified muscle cells serve as the initiation point for the heart's electrical system. These cells, located in a 0.8 × 0.08 inch (2 cm × 2 mm) area in the atrial wall near the entrance of the superior vena cava, are collectively known as the **sinoatrial node** (SA node), or the **pacemaker**. These cells are electrically connected, so when one "fires"—delivers an electrical impulse—they all do. The pacemaker fires spontaneously and needs no nervous system input to continue to deliver regular electrical impulses. In fact, the heart will continue to beat for a while even if it is completely removed from the body. On the other hand, the nervous system is important in that it can override the pacemaker's regular firing rate and either slow it down or quicken it. The heart also has backup regions that can take over if the pacemaker is compromised. These regions, called **ectopic pacemakers**, are capable of initiating the heartbeat when necessary.

When the pacemaker fires, the electrical impulse spreads at a rate of about 3.29 feet (1 m) per second to the left and right atria and causes them to contract. As noted earlier, the fibers in each cardiac muscle cell are connected to fibers in adjoining cells. This connection allows the cells to contract nearly in unison. As the atria contract, the blood flows past the respective valves and into the left or right ventricle. At the bottom of the septum dividing the atria is a small group of cells and connective tissue known as the **atrioventricular node**, or **AV node**. The AV node is the only conducting path through the annulus fibrosus that divides the atria and ventricles. This node gets the electrical impulse from the pacemaker at about the same time as the atria do, but forwards it much more slowly to the ventricle, resulting in a delay of about a tenth of a second. This allows time for the atria to contract and squeeze the blood into the ventricles. The AV node then relays the impulse not directly to the ventricle, but to two structures. The first of the two is the **bundle of His** (pronounced "hiss"), a thick conductive tract that transmits the signal to a mesh of modified muscle fibers, called the **Purkinje fibers** (pronounced "purr-kin-gee"), in the base of the ventricle wall. It is these fibers that pass the impulse to the ventricle (at a speed of 5.5 feet [1.6 meters] per second!).

The ventricle contracts beginning at the bottom. As the wave of contraction progresses upward, it efficiently forces the blood upward to the exit valves.

Normal Heart Function Variations

Although the heart beats about 70–80 times per minute in a resting adult, that rate can vary considerably depending on whether the person is sitting down or standing up, walking or running, relaxed or under stress—activity of nearly any sort, as well as various medical conditions and medications, can cause the heart to speed up or slow down. In addition, stroke volume (the amount ejected from the left ventricle to the aorta with each beat) can also change to meet demands.

This formula is used to obtain the cardiac output, which is the volume of blood pumped over a certain period of time (typically the amount of blood ejected by one ventricle in one minute):

$$(\text{stroke volume}) \times (\text{heart rate}) = \text{cardiac output}$$

On average, the stroke volume of an adult at rest is about 0.08 quarts (75 ml). When multiplied by the average male's heart rate of 70 beats per minute (a female's is about 78 beats per minute), the cardiac output comes to 5.5 quarts (5.2 liters) per minute. Cardiac output varies greatly among individuals, and while 5.5 quarts per minute is the approximate overall average, a healthy individual can fall within the range of about 4.2–7.4 quarts (4–7 liters) per minute. Generally, an individual's cardiac output at rest is 3.2 quarts (3 liters) per minute for every 3.3 square feet (or 1 square meter) of body surface area. Using that calculation, an adult weighing 150 pounds (68 kg) and having a body surface area of about 5.9 square feet (1.8 square meters) would have a cardiac output of 5.4 quarts (5.1 liters) a minute.

That cardiac output is distributed to the body more or less on the following principle: the higher the metabolic rate of the tissue, the higher the percentage of the cardiac output it receives. Muscles use about a fifth of the oxygen a person breathes, so they receive about a fifth of the blood. Two of the most notable exceptions to the rule are the kidneys and the heart. The kidneys receive as much blood as the muscles, even though the kidneys use less than a third of the oxygen. High blood flow to the kidneys is necessary because these organs are the body's blood filters, removing waste products and excess water. The heart, on the other hand, receives proportionately less blood than it should, if figured according to the rule. It compensates, however, by drawing more oxygen from its limited supply than other organs do, thanks to its particularly dense capillary network.

For the most part, the autonomic nervous system rules the variations of the heart rate, which in turn affects cardiac output. The autonomic system, which controls involuntary activities, has two major divisions: the sympathetic nervous system and the parasympathetic nervous system. (See Chapter 8 on the nervous system for more information.) The former stimulates the pacemaker and boosts the heart rate, while the latter inhibits the pacemaker and lowers the heart rate. The sympathetic and parasympathetic systems also have opposite effects on the arteries, with the former causing them to contract and the latter causing them to dilate. In effect, the sympathetic side prepares a person to respond to stressful situations by heightening blood flow, and the parasympathetic (also known as vagal) side brings a person back to normal.

Besides the autonomic nervous system, other factors can influence cardiac output, including various hormones, baro- and chemoreceptors, and the fitness level of the individual.

Receptors

Every time the heart beats, the blood pressure increases in the aorta and in the carotid sinuses, which are swellings or expansions at the base of the internal carotid arteries. When the heart's contraction ends, the pressure decreases. Detectors in the walls of major arteries perceive those pressure changes by sensing the tension in the vessel walls. The detectors, neurons called baroreceptors, pass this information to the parasympathetic nervous system, which responds as necessary by lowering the heart rate and decreasing its contractility, which together cause a dip in blood pressure. The sympathetic nervous system also heeds the call by reducing arterial tone and making arteries more elastic, which also serves to decrease blood pressure. The major baroreceptors are the aortic baroreceptors that keep track of blood pressure in the ascending aorta, and the carotid sinus baroreceptors in the neck that track blood flow to the brain. In addition, atrial baroreceptors at the venae cavae and right atrium check blood pressure as blood enters the heart from the venous system. When more blood is entering the heart than is being pumped out, the body rectifies the imbalance by heightening cardiac output until the incoming and outgoing blood are equalized again. In summary, the baroreceptor system is an effective means for evening out the short-term spikes and dips in blood pressure.

Besides pressure detectors, the body has chemoreceptors to monitor the levels of oxygen and carbon dioxide in the blood. These receptors, located near the carotid sinus and the aortic arch, also detect acidity, or pH, levels. The amount of carbon dioxide in the blood can alter its acidity level, because carbon dioxide dissolves in the water of the blood plasma and makes carbonic acid. The more acidic the blood, the lower the pH. When chemoreceptors sense a rise in carbon dioxide levels, a fall in oxygen levels, or a drop in pH, they trigger a hike in cardiac output, which leads to higher arterial blood pressure. Other chemoreceptors in the medulla oblongata (lower brain stem) keep track of blood composition going to the brain and respond to reduced oxygen by initiating the dilation of cerebral vessels while constricting vessels to other parts of the body. The body, then, works to maintain blood flow to the brain at the expense of other organs and systems.

Blood-Demanding Organs and Circulation-Related Systems

Blood is almost everywhere in the human body. It flows to all of the tissues, moves in and out of organs, and participates in all sorts of bodily functions. This section will introduce some of the organs that put a high demand on blood flow, as well as the lymphatic system that has a close bond with circulation.

Digestion

The main arteries feeding the stomach include the celiac, gastric, gastroduodenal, and gastroepiploic arteries. The **celiac artery**, or celiac trunk, stems from the abdominal aorta, which is the portion of the aorta that lies in the abdomen (as opposed to the thoracic aorta that runs through the chest cavity). The celiac artery branches into the left gastric, common hepatic, and splenic arteries. For the digestive system, the left gastric artery supplies blood to the stomach and the lower part of the esophagus, which is the feeding tube extending from the pharynx at the back of the mouth to the stomach. The **gastroduodenal artery** separates from the common hepatic artery, which itself continues on to the liver as the hepatic artery. The gastroduodenal artery, the right gastroepiploic artery that branches from it, and the left gastroepiploic artery that arises from the splenic artery all provide blood to the stomach and duodenum (the first part of the small intestine). The right gastric artery arises not directly from the celiac artery like the left gastric artery does, but from the hepatic artery proper. It eventually connects with the left gastric artery and supplies blood to the stomach.

Blood to the colon comes from the left, middle, and right colic arteries, all of which branch from either the inferior or superior mesenteric arteries. Both mesenteric arteries separate from the abdominal aorta. The mesenteric arteries also supply blood to the rest of the intestinal system through the intestinal, ileocolic, and other arteries.

How does the food a person eats wind up in the blood? The answer lies in tiny outgrowths, called villi (the singular is villus), that line the inside of the wall of the small intestine (Figure 2.8). These villi are, in turn, coated with even smaller outgrowths, called **microvilli**. By the time digested food reaches the small intestine, it has already been broken down into molecules small enough to cross the capillary walls. Each of the villi has a set of capillaries, thus providing ample opportunity for the uptake of food from inside the small intestine into the blood system. Although the work of the villi might seem inconsequential, they are so numerous throughout the small intestine and are covered by such a large number of microvilli that they actually increase the surface area of the interior small intestine by about 600 times. This puts the blood into close contact with much of the digested material, or chyme, before it makes the approximately two-hour journey through the small intestine.

The mesenteric veins and other feeder arteries serve as the exit route for blood from the intestines and colon. Gastric and gastroepiploic veins drain the stomach into a number of other veins that ultimately unite—along with the mesenteric veins—at the large portal vein. Thus, nutrient-laden

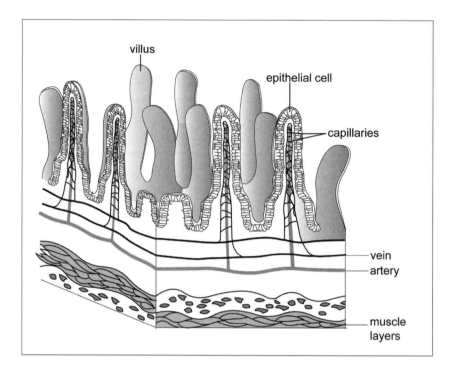

Figure 2.8 Villi. A multitude of tiny outgrowths, called villi, line the inside of the wall of the small intestine. These villi are, in turn, coated with even smaller outgrowths, known as microvilli (not shown). Each villus has its own set of capillaries, thus providing ample opportunity for the uptake of food from inside the small intestine into the blood system. (Sandy Windelspecht/Ricochet Productions)

blood leaves the digestive system through the portal vein, but before it begins its return trip to the heart, it heads to the liver.

Liver and Hepatic Circulation

The liver is a large, broad organ (sometimes referred to as a gland) that sits below the lungs and diaphragm, but above the stomach. Measuring about 8–9 inches (20–23 centimeters) long and 6–7 inches (15–18 centimeters) high, this two-lobed, wedge-shaped organ plays an important part in the use and storage of food energy.

About a quarter of the entire cardiac output passes through the liver, but unlike many organs, the liver gets only 25–28 percent of this blood directly from the aorta or its primary branches. That amount, however, is enough to support the organ and its functions. Most of the remaining 72–75 percent of the blood arriving at the liver has already been through the digestive system, where it picked up nutrients from food. This indirect type of blood distribution is termed **in-series blood circulation** because the blood goes from one organ to another in a series. It is also called **portal circulation**.

As soon as the blood arrives in the liver from the portal vein, the liver gets to work removing the food molecules, including fats, the amino acids from proteins, and sugars, and converts them into a carbohydrate called glycogen for storage as a reserve energy supply for the body to use later as the need arises. The sugars, called glucose, switch readily to glycogen, but the fats and amino acids must first be converted to glucose before they make the change to glycogen. Because the liver is capable of converting glycogen back into glucose, it serves as a regulator of the **blood-sugar level**.

The liver is also one of the body's lines of defense against drugs, alcohol, poisons, pollutants, and other chemical threats. Just as it filters out sugar, the liver selectively removes these substances via the urine or the bile. In addition to its responsibilities in metabolism and detoxification, the liver makes other compounds, including cholesterol.

Blood exits the liver through the hepatic vein that joins the inferior vena cava for the return trip to the heart.

Kidneys and Renal Circulation

As described previously, the kidneys are the body's blood filters and the site where urine is produced. The two kidneys, shaped appropriately like kidney beans, are each about 4 inches (10 centimeters) long and 2 inches (5 centimeters) wide, and are located in about the middle of the back, with one on either side of the backbone.

Blood is approximately 80 percent water, so the body requires a competent system to maintain a proper water balance. The kidneys provide that service, removing excess water and also filtering out waste products. The water and waste products, including soluble materials like salts, are all eliminated via the formation of urine, which exits each kidney through a long tube, called a **ureter**, that flows into the bladder.

The filtering process begins when blood—about a fifth of the total cardiac output—arrives at each kidney from a separate renal artery. **Interlobar arteries** branch from the renal artery to disperse the blood throughout the kidney and to networks of capillaries, called glomeruli (glomerulus is the singular). The glomeruli come into contact with the kidney's filtering units, called **nephrons**, which are each composed of a twisting epithelial tube. On one end, the epithelial tube eventually empties into the ureter, and on the other end, terminates in a bulb. The bulb, known as a **Bowman's capsule**, surrounds the glomeruli and provides an efficient transfer site for water and waste products to move from the blood to the urinary system. Blood leaves the Bowman's capsule in one vessel, but that vessel quickly branches into a second set of capillaries. This time, the capillaries form a web that weaves around the nephron tubules, which participate in returning much of the water and many of the solutes to the blood through the capillaries. Excess water and solutes continue through the nephron to the ureter. Blood exits each kidney through a renal vein.

In a single day, a single person's kidneys can filter out some 40 gallons (151 liters) of water. Reabsorption returns about 39.6 gallons (150 liters). The difference, 33.8 ounces (1 liter), is about the same as the amount of water ingested by that person in a 24-hour period. Thus, the body neither adds nor loses water in a typical day.

Right and left renal veins drain the kidneys. These veins also transport blood from the adrenal glands that sit atop the kidneys. (See Chapter 12 for more information on the kidneys.)

Spleen

The spleen is a mainly red, heart-sized organ that sits to the left and below the stomach. As noted in the section on the digestive system, one of the branches from the celiac artery is the splenic artery. This artery supplies blood not only to the stomach, but also to the spleen.

This organ has two primary functions: assisting in the removal of foreign materials and aging red blood cells from circulation, and storing red blood cells and platelets.

To spot and eliminate the threat from invading organisms, the spleen relies on the immune system. The organ contains many branching blood vessels that are enveloped with B lymphocytes and T lymphocytes (B cells and T cells). The T cells have a mission of surveillance and scan the slowly flowing blood for the foreigners, like bacteria. They report invaders to the B cells. Any B cell that has previously encountered the bacterium rapidly multiplies and then begins producing antibodies to weaken or destroy the invader. The spleen is also armed with many **macrophages**, the white blood cells that ingest and digest bacteria, other foreign organisms, platelets, and old or deformed red blood cells.

In addition, the spleen can swell to hold blood in storage. This ability provides a reserve supply that can be tapped when necessary. Besides the filtration and storage functions, the spleen has another job in the human embryo: It makes red blood cells, a function that shifts to the bone marrow after birth.

Cerebral Circulation and the Blood-Brain Barrier

As discussed previously, the maintenance of blood flow to the brain is one of the circulatory system's highest priorities. This is because the brain is key to so many vital physiological processes. The main arteries to the head include the left common carotid that branches directly off of the aortic arch, and the right common carotid that branches indirectly from the aortic arch by way of a short brachiocephalic (also called innominate) artery. The two carotid arteries traverse almost straight up, branching again at about chin level into internal and external carotids. Other arteries feeding the head include the **vertebral arteries**. These two arteries, one on each side of the neck, arise from the right and left subclavian arteries that divert from the aorta. The vertebral arteries unite at the **basilar artery**, and this artery joins with other cerebral arteries to form what is known as the **circle of Willis**.

Blood supply to the brain comes from numerous major arteries, as well as smaller, branching arteries and arterioles. For example, the internal carotid artery supplies blood to the anterior brain, and one of its branches, called the anterior cerebral artery, feeds the cerebrum. The cerebrum comprises the two large hemispheres of the brain. In some cases, several arteries may supply the same area of the brain. The cerebrum also receives blood from the posterior cerebral artery, which derives from the basilar artery that arises from the vertebral artery. Blood flow to the brain, then, emanates from numerous arteries and arterioles.

Similarly, blood drains from the brain via a large number of venules and veins that empty into several large veins. These include the vertebral vein, and the internal and external jugular veins. The internal jugular is by far the largest of the three, and runs almost down the middle of the neck. It serves as the primary collector for deoxygenated blood, which it delivers to the subclavian vein and eventually to the superior vena cava for its return to the heart.

Although the brain accounts for only 2 percent of an average person's weight, it demands approximately one-fifth to one-sixth of the cardiac output, or a flow of about 0.74 quarts (700 ml) per minute in an adult. The kidney and liver both require a high percentage of the cardiac output for such maintenance functions as water filtration or waste removal. The brain, on the other hand, needs the large quantity of blood for the oxygen. This heightened demand stems from the extreme rate of oxidative metabolism seen in the nerve cells, or neurons, of the brain. These neurons, collectively called gray matter, account for about 40 percent of the brain and use almost 20 percent of the oxygen that a resting person breathes. Just 1 square millimeter of gray matter can have up to 4,000 capillaries. In addition, the gray matter is damaged very quickly if the oxygen supply drops off. Fainting is one of the body's responses to this condition of hypoxia (low oxygen levels). This brings the head closer to the ground, which serves to take advantage of gravity and allow blood to flow more readily to the brain. Continued hypoxia lasting more than a couple of minutes can cause permanent brain damage.

Blood-Brain Barrier

The blood-brain barrier is actually a collection of tightly enmeshed cells and other obstacles that effectively serve as a boundary, allowing only oxygen, certain nutrients, and a few other items to pass into brain tissues via the blood (Figure 2.9). Researchers Paul Ehrlich and Edwin Goldman saw in the late 1800s to early 1900s that dyes injected into the brain and

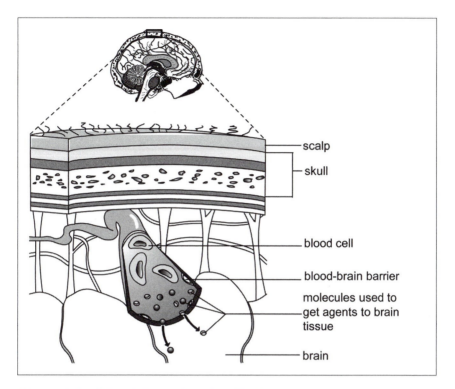

scalp

skull

blood cell

blood-brain barrier

molecules used to get agents to brain tissue

brain

Figure 2.9 Blood-Brain Barrier. The blood-brain barrier is a protective boundary that allows only certain gases, nutrients, and other materials to pass from the blood into the brain tissues. In some cases, special molecules help to transport materials across the barrier. Scientists are now learning more about these transportation avenues and other ways to circumvent the barrier to allow beneficial materials, like drugs, to enter the brain. They are also studying why such dangerous entities as bacterial meningitis are able to circumvent the barrier. (Sandy Windelspecht/ Ricochet Productions)

cerebrospinal system would only color blood there, and dyes injected elsewhere would color all the blood except that in the brain and cerebrospinal system. A barrier existed that selectively prohibited various toxins and other substances from entering the brain. This is now known to protect the brain and its myriad nerve cells from diseases and chemicals that might impair its function. It also, however, prevents many helpful medications from penetrating into the brain, and researchers have been studying how to circumvent the barrier for several decades.

Summary

This chapter provides a brief overview of the very complex circulatory system. The system's primary organ is the heart, which continuously forces blood into and out of the body's blood vessels. The blood is another vital component of this system, and is responsible for transporting important materials throughout the body, including oxygen to the lungs and the heart. The blood is made up of red and white cells, as well as plasma and platelets, all of which have important roles in ensuring the blood performs its duties. Another important component of the circulatory system consists of the blood vessels, which are tissues that transport the blood to where it is needed. While the vessels allow the blood to travel throughout the body, the capillary system facilitates the exchange between the blood and various tissues, including the vessels. Since all of the body's systems are interconnected and depend on one another, it is not surprising that organs from other systems rely on the blood and circular system to function. These include the liver, stomach, and other organs of the digestive system; the kidneys and renal organs; as well as the brain—the primary organ of the nervous system.

3

The Digestive System

Michael Windelspecht

Interesting Facts

- The salivary glands can produce up to 1.5 quarts (1,500 milliliters) of saliva daily.

- Food remains in the esophagus for as little as five seconds before entering the stomach.

- The human stomach can hold as much as 2.1 quarts (2 liters) of food.

- The stomach produces 2.12 quarts (2 liters) of gastric juice daily.

- Gastric juice is 100,000 times more acidic than water and has about the same acidity as battery acid.

- The small intestine in an adult can reach 3.28 yards (3 meters) in length, while the large intestine is only about 1.64 yards (1.5 meters) long.

- Every square millimeter of the small intestine can contain 40 villi and 200 million microvilli.

- Food remains in the small intestine for three to five hours on average, during which most nutrients are removed.

- The intestines receive over 10 quarts (9 liters) of water daily, of which almost 95 percent is recycled back into the body.

- As little as 10–15 percent of the iron we eat in food is absorbed into the human body.

Chapter Highlights

- The mechanics of digestion

- Different types of energy nutrients

- Vitamins, minerals, and water

- Upper gastrointestinal tract (oral cavity, esophagus, and stomach)

- Lower gastrointestinal tract (small intestine and colon)

- Large intestine

- Accessory organs (salivary glands, liver, gall bladder, pancreas)

Words to Watch For

Alkaline

Amphipathic
  molecules

Antioxidants

Autocatalytic process

Bilirubin

Binucleate cells

Bioavailability

Bolus

Buccal cavity

Cholesterol

Deglutition

Dehydration
  synthesis

Dentin

Diaphragm

Diploid cells

Electrolytes

Epithelial cells

Exocrine gland

Facilitated diffusion

Hemoglobin

Hepatic portal system

High-density lipopro-
  teins (HDL)

Hormones

Hydrolysis

Hydrophilic

Hydrophobic

Ions

Kilocalorie

Lacteals

Lipoproteins

Low-density lipopro-
  teins (LDL)

Lumen

Mesentery

Metabolism

Organic molecules

Peristaltic action

Pharynx

Phospholipids

Polyploid cells

Introduction

One of the unifying characteristics of all living organisms is their ability to process nutrients from the environment into the chemical compounds found within the cells. This processing of nutrients is commonly called **metabolism**. Plants, animals, fungi, and bacteria have all evolved different strategies for supplying the energy and chemical needs of the organism. Simply stated, digestion is the breakdown of food particles into their fundamental building blocks. As heterotrophic organisms, or those that rely on others as a source of energy, animals have evolved a wide range of digestive systems to accommodate their environmental needs. The internal body plan of an animal species is frequently defined by its digestive system. The purpose of this chapter is to examine the structure and function of the digestive systems in the species Homo sapiens, or humans.

The digestive system is not a stand-alone system. The primary digestive organs—the mouth, esophagus, stomach, and small and large intestine—not only interact to some degree with each other, but also receive signals from other organs of the body. The accessory organs—the liver, pancreas, salivary glands, and gall bladder—supply chemicals necessary for the nutrient processing. The liver and pancreas are active with other systems of the body as well, such as the endocrine and circulatory systems. The digestive system is partially under the control of the nervous system, but also is influenced by the hormones secreted by the endocrine system. Because nutrients absorbed by the digestive system must be transported throughout the body, the gastrointestinal tract interacts with the circulatory and lymphatic system for the transport of water-soluble and fat-soluble nutrients, respectively. Finally, the urinary system removes some of the waste products of nutrient metabolism by the liver.

The purpose of digestion is to process food by breaking the chemical bonds that hold the nutrients together. This is necessary so that the body has an adequate source of energy for daily activity, as well as materials for the construction of new cells and tissues. Since these nutrients arrive in the digestive system as the tissues of previously living organisms, they are rarely in the precise molecular structure needed by a human body. For example, the blood of cows and chickens has evolved over time to meet the precise metabolic needs of the organism. When the tissues of these animals are consumed, our bodies must chemically alter the proteins and other nutrients found in the animal's blood to form human blood proteins such as hemoglobin. As is the case with almost all the nutrients (with the exception of water, minerals, and some vitamins), the body breaks down the nutrient into its fundamental building blocks, transports the digested nutrient into circulatory and lymphatic systems, and eventually uses these nutrients in the cells of the body for either energy or metabolic processes. Before proceeding into how these reactions occur, it is first necessary to provide an overview of the digestion process, as well as to discuss the basic nutrient classes.

Digestion at the Cellular and Molecular Level

When the term digestion is mentioned, it is natural to think about the actions of the mouth, stomach, and small intestine in the processing of food for energy. While these actions are no doubt important in the breakdown of food, they actually are the result of complex processing mechanisms at the cellular level. This chapter will examine the physiology of the digestive system at the organ level. However, to effectively understand the structure and function of the digestive system, we must first understand the cellular and molecular basis of nutrient processing.

The purpose of digestion is to process food by breaking the chemical bonds that hold the nutrients together. This is necessary so that the body has an adequate source of energy for daily activity, as well as materials for the construction of new cells and tissues. Since these nutrients arrive in the digestive system as the tissues of previously living organisms, they are rarely in the precise molecular structure needed by a human body. For example, the blood of cows and chickens has evolved over time to meet the precise metabolic needs of the organism. When the tissues of these animals are consumed, our bodies must chemically alter the proteins and other nutrients found in the animal's blood to form human blood proteins such as **hemoglobin**. As is the case with almost all the nutrients (with the exception of water, minerals, and some vitamins), the body breaks down the nutrient into its fundamental building blocks, transports the digested nutrient into circulatory and lymphatic systems, and eventually uses these nutrients in the cells of the body for either energy or metabolic processes.

Classes of Nutrients

There are six general classes of nutrients: carbohydrates, fats, proteins, water, vitamins, and minerals. Carbohydrates, fats, and proteins are characterized as energy nutrients. These **organic (or carbon-containing) molecules** are responsible for providing our bodies with the majority of the energy needed for daily metabolic reactions. This does not mean that the remaining nutrient classes are not important in energy reactions within the body. In fact, many of these, such as some of the B vitamins and water, are crucial to the efficient operation of the energy pathways. However, our bodies do not get energy from these nutrients directly.

Carbohydrates, proteins, and fats all contain energy in the carbon-carbon bonds of their molecules. The energy of these bonds is measured in a unit of heat measurement called the calorie. A calorie is the amount of energy required to raise 1 gram of pure water by 1 degree Celsius at sea level. However, this is a relatively small unit of measurement, and thus for nutritional analysis the term **kilocalorie** (1,000 calories) is frequently used. When one examines the ingredient label of a prepared food, such as a soft drink, the listed calorie value is actually in kilocalories, also called kcals. Organic molecules contain a large number of carbon-carbon bonds, and are therefore an excellent source of metabolic energy.

Energy from Energy Nutrients

Cells have a variety of mechanisms for releasing the energy contained within the carbon-carbon bonds of organic molecules. Some cells are **anaerobic** and can obtain small amounts of energy without the assistance of oxygen. However, the majority of the cells of the body utilize a complex metabolic pathway called aerobic respiration. Aerobic respiration consists of three main series of reactions: **glycolysis**, the Krebs cycle, and the electron transport chain (ETC). The Krebs cycle and ETC occur in the **mitochondria** of the cell and use oxygen to regenerate a cellular energy molecule called adenosine triphosphate (ATP). In the Krebs cycle, the carbon-carbon bonds are broken and a small amount of ATP is generated. The remaining carbon is combined with oxygen to form carbon dioxide, a waste product. The details of aerobic respiration will be covered in greater detail in Chapter 10 on the respiratory system.

TABLE 3.1
The Energy Nutrients

Energy nutrient	Monomer	Polymer	General enzyme	Energy per gram
Carbohydrates	Monosaccharides	Polysaccharides	Amylases	4 kcal
Proteins	Amino acids	Proteins	Proteases	4 kcal
Fats and lipids	N/A	N/A	Lipases	9 kcal

This table lists some important facts regarding the energy nutrients. The "N/A" under fats and lipids reflects the fact that these molecules do not form complex structures in the same manner as carbohydrates and proteins. The energy per gram is an approximation and varies depending on the metabolic properties of the cell and individual.

The aerobic respiration pathways are capable of utilizing most organic molecules as an energy source. However, in order for the nutrients to enter the pathway, they must first be broken down into their fundamental building blocks. Proteins and carbohydrates are actually long repetitive chains of individual building blocks called monomers. These monomers are linked by chemical bonds into long polymers (see Table 3.1), which first must be broken down. Water is used to break the bonds linking the monomers in a process called **hydrolysis**. Cells use the reverse of this process, called **dehydration synthesis** or condensation reactions, to form more complex molecules from the monomers.

Carbohydrates

In the study of nutrition, carbohydrates are frequently abbreviated as CHO, which reflects the fact that this nutrient class contains the elements carbon, hydrogen, and oxygen. All carbohydrates possess carbon, hydrogen, and oxygen in a 1:2:1 ratio, respectively. For example, the molecular formula for glucose is $C_6H_{12}O_6$. Carbohydrates are the short-term energy molecules of the human body and are the preferred fuel of the aerobic respiration pathways.

All carbohydrates are made up of one of three building blocks, or monosaccharides. The most common of these is glucose, with the other two being fructose (the monosaccharide associated with the sweet taste) and galactose (sometimes called a milk sugar). Glucose is the preferred energy molecule for aerobic respiration, and the human digestive system is well adapted to extracting this nutrient from foods and delivering it to the cells. All of the monosaccharides are water-soluble and easily transported by the circulatory system.

Monosaccharides are linked together in pairs by chemical bonds to form the disaccharides. All disaccharides contain at least one glucose unit in their structure. There are three different disaccharides: maltose (glucose-glucose), sucrose (glucose-fructose), and lactose (glucose-galactose). Together, the monosaccharides and disaccharides all are commonly called the simple sugars.

Complex carbohydrates, or the polysaccharides, are composed of long chains of glucose units. The different classes of polysaccharides vary in the physical structure of the chemical bonds that link the glucose units. In some cases, such as starch, the chemical bonds are easily digested by the human digestive system and thus provide a useful source of glucose for energy. However, a slight change in the configuration of the chemical bonds between the glucose units makes the bonds inaccessible by human digestive enzymes.

These molecules are called fibers, and even though they are not digestible by human enzymes, they play an increasingly important role in human digestion. Since they are basically indigestible, fibers provide bulk to food. This bulk helps move materials through the system and provides resistance to muscles of the gastrointestinal tract. This resistance acts as a form of workout for the muscles, which keeps them strong, allowing for the efficient movement of food in the system. Fibers exist in one of two general categories: soluble and insoluble. The soluble fibers, which are readily dissolved in water, are found primarily in fruits. These fibers slow down the movement of food through the gastrointestinal tract as well as the absorption of glucose. In contrast, the insoluble fibers, which are found in bran material and whole grains, increase the rate at which material is moved through the gastrointestinal tract, as well as provide bulk to the fecal material.

Fats and Lipids

While the carbohydrates are regarded as short-term energy sources for the human body, the fats and lipids are involved with more long-term energy processes within the body. There are exceptions to this, but in general the fats and lipids take more time to process by the digestive system and are associated with developing the long-term energy stores of the body. Technically, the term lipid is used to represent the entire class of these molecules, with the term fat primarily being reserved for a group of lipids called the triglycerides. However, frequently in nutritional analysis and on consumer products, the terms are used interchangeably. In this volume, the term fat will be reserved for the triglycerides, with lipids indicating the entire class of molecules. Two major classes of lipids are of interest in understanding the physiology of the digestive system of humans: the triglycerides and the sterols. A third class, the **phospholipids**, plays an important role in the structure of cell membranes.

For the most part, the lipids are **hydrophobic** molecules, meaning that they do not dissolve readily in water. Because the digestive system is a water environment, as is the circulatory system, this physical characteristic of the lipids means that the digestive system will have to handle the lipids differently than most other nutrients. Chemical secretions such as bile, and specialized proteins called the **lipoproteins**, will assist the processing and transport of these important energy nutrients.

The triglycerides make up the majority (95 percent) of the lipids in food. The structure of these molecules, with its high number of carbon-carbon bonds, makes them an excellent source of energy for aerobic respiration. The long chains, called fatty acids, can vary in length and their degree of saturation. It is these chains that make the triglycerides hydrophobic. The level of saturation of the fatty acid chains has also been associated with human health. Saturated fats, found typically in animal products, are known to increase the risk of heart disease, while the unsaturated fats of plant products produce less of a risk. While the words triglyceride and fat have a negative connotation in today's society, in fact they are necessary and useful molecules in human metabolism when they are consumed in correct quantities. Some fats, called the omega-3 and omega-6 fatty acids due to their structure, actually help regulate the lipid biochemistry of the blood. Fats and lipids also provide insulation for the body. Because the fats give texture to food, release a pleasing aroma when cooked, and provide a fullness to the meal, they can be easily over-consumed during eating.

A second major class of lipids is the sterols, of which the most common is called **cholesterol**. As is the case with the triglycerides, the word cholesterol does not have a positive image in today's society. However, just like the triglycerides, cholesterol is an important molecule for our bodies. It serves as the starting material for the manufacture of important **hormones** such as testosterone and estrogen, is a component of the membranes of our cells, used by the liver to manufacture bile, and is the starting material for the synthesis of vitamin D in our bodies. In fact, cholesterol is such an important molecule to our bodies that our liver has the capability of manufacturing all of the body's daily requirements of cholesterol.

Like the triglycerides, cholesterol is a hydrophobic molecule, and thus the body can have some problems moving it around. To remedy this, cholesterol as well as triglycerides are packaged into a group of special transport molecules called lipoproteins. Think of these lipoproteins as balloons. When the balloons are empty of cholesterol, they are compact and small and are called **high-density lipoproteins (HDL)**. When there is an abundance of cholesterol in the system, the balloons are full, and they are called **low-density lipoproteins (LDL)**. Unfortunately, these lipoproteins have been given the names "good" (HDL) and "bad" (LDL) cholesterol, but that really is not correct. Because the body has the ability to manufacture all of its cholesterol, an overabundance in foods, called dietary cholesterol, will fill the balloons creating LDLs. A diet low in fat and cholesterol will leave the balloons empty, which are the HDLs. Scientists and nutritionists are still debating the effects of dietary cholesterol on the body. The amount of cholesterol does not directly influence the operation of the digestive system (although the foods that cholesterol is associated with do), but it does affect the health of the circulatory system.

There are additional classes of lipids, such as the waxes and phospholipids. Although important to the overall operation of the human body, they have little influence on the physiology of the digestive system and are processed in the same manner as the triglycerides and sterols.

Proteins

While the body primarily uses lipids and carbohydrates for energy, proteins are involved in a wide variety of functions other than supplying energy. To put it simply, proteins are the working molecules of the cell and, as such, are involved in the majority of all cellular functions. Proteins may have structural functions, such as those found in muscles; others work as signaling molecules in the nervous system or as hormones for the endocrine system. This list of protein functions in the human body is almost too numerous to mention, but a specialized group of proteins, called the enzymes, are an important component of the digestive system and are covered in the next section.

The building blocks of proteins are the **amino acids**. Twenty different amino acids are needed to construct the proteins of the human body. Each of these amino acids has a slightly different molecular structure, which gives each of them a unique chemical characteristic. Within each cell, using instructions from the genetic material (deoxyribonucleic acid, or DNA), amino acids are linked together by the process of dehydration synthesis to form proteins. Like other molecules, these bonds are broken by the process of hydrolysis. However, unlike the carbohydrates, when the amino acid chains are formed, they fold into complex three-dimensional structures that inhibit digestion in the body. Furthermore, the peptide bonds that hold the amino acids together are exceptionally strong, thus requiring the assistance of enzymes to break them down.

Because the proteins in food are the product of the original organism's body, and were constructed by the organism for a specific purpose, few proteins that are brought into the human digestive system are usable in their current form. Instead of absorbing whole proteins into the body, the role of the digestive system is to break down the protein bonds into their amino acid building blocks. The individual amino acids are then absorbed into the body and used as raw materials for the building of human-specific proteins. The digestion and absorption of proteins will be covered in more detail later in this chapter.

Enzymes

For the most part, the metabolic reactions of the body, including the dehydration synthesis and hydrolytic reactions mentioned previously, do not occur spontaneously. Instead, they require a catalyst to accelerate the rate of the reaction to a point that is efficient for the cells of the body. These catalysts are called enzymes. While the human body has a wide array of enzymes, which control everything from the operation of the nervous system to the process of cell division, they all share some common characteristics.

First, the vast majority of all enzymes are proteins. The three-dimensional shape of enzymatic proteins enables them to interact with other molecules. Second, enzymes are very specific to the molecules, or substrates, with which they interact. Third, enzymes all serve to increase the efficiency of metabolic reactions by lowering the amount of energy needed to initiate the reaction. Finally, enzymes themselves are not consumed or destroyed during the course of an enzymatic reaction, allowing them to be reused over and over again for the same process.

The activity of an enzyme may be regulated by a variety of mechanisms. First, enzymes all have a specific environment in which they are the most efficient. The temperature and pH (or acid/base level) of the enzyme's environment act as a switch to regulate its activity. Since within the digestive system of humans, the temperature remains a relatively constant 98.6°F (37°C), digestive enzymes are primarily regulated by the pH of their environment. The level of compartmentalization in the human digestive system helps to establish zones of enzyme activity. Throughout this chapter, we will examine how the stomach and small intestine regulate the pH of their environments to control enzyme activity.

The digestive system utilizes a large number of enzymes to break down the nutrients within food into units small enough to be transported by the circulatory or lymphatic system. In this volume, we will usually refer to the enzymes by their general function. For example, enzymes that assist in the processing of lipids are called lipases, and those that process proteins are called proteases. The prefix "amyl-" means sugar. Other enzymes and their mechanism of regulation will be discussed in the next several chapters.

Vitamins, Minerals, and Water

The action of the digestive system is not confined solely to the processing of the energy nutrients. The human body requires a daily input of other nutrients to meet its metabolic requirements. The processing of vitamins, minerals, and water differs from that of the energy nutrients in that these nutrients are usually not broken down by the digestive system, but rather are absorbed intact and then transported by the circulatory system to the other systems of the body. While is it beyond the scope of this encyclopedia to describe all of the vitamins and minerals, some basic characteristics of these nutrients are described in the following paragraphs so as to provide an overview of how these nutrients interact with the digestive system. The bibliography of this encyclopedia provides a list of useful sources for additional information on specific nutrients.

Vitamins

Vitamins are similar to the energy nutrients in that they are organic molecules, but differ in the fact that the body does not get energy directly from these molecules. Instead, vitamins serve as enzyme assistants, or coenzymes. Some vitamins, specifically the B-complex vitamins, are directly involved in the processing of energy nutrients, specifically lipids and carbohydrates. Certain vitamins serve as protectors of the delicate cellular machinery. These are called the **antioxidants** and are best represented by vitamins C and E. Others aid in the vision pathways (vitamin A), or in the building of healthy bones (vitamins D and A). Nutritionists divide the vitamins into two groups based upon how they interact with the body. The first are the water-soluble vitamins, a group that consists of vitamin C and the B vitamins. These vitamins are readily absorbed by the digestive system and, with a few exceptions, do not require special processing. The other class, known as the fat-soluble vitamins (vitamins A, D, E, and K), are frequently treated in the same manner as the triglycerides, meaning that they are packaged into specialized lipoproteins and transported by the lymphatic system. In general, both classes are required in relatively small quantities (micrograms or less) daily by the body.

There are two vitamins that we will focus on in some detail. The first is vitamin D, which is produced by the body using the cholesterol in the skin as a starting material. When the skin is exposed to sunlight, specifically ultraviolet radiation, the chemical structure of the cholesterol is modified to create a precursor of vitamin D. This chemical is then transported to the liver and adrenal glands for additional processing. Vitamin D acts like a hormone, in that it regulates the calcium absorption properties of calcium in the small intestine.

The second vitamin of interest in the study of the digestive system's physiology is vitamin K. Vitamin K is a vitamin that is involved in a wide variety of body functions, most notably the clotting response of the blood.

Some vitamin K is produced by the naturally existing bacteria of the large intestine, or colon. As with many of the nutrients, there is a significant amount of misinformation in the popular media regarding the ability of some vitamins to prevent disease, enhance performance, increase memory, and so on.

Minerals

Minerals are inorganic nutrients that play an important role in the regulation of many of the body's metabolic functions. Like the vitamins, many minerals function as assistants to metabolic pathways. Still others help regulate body fluid levels, and some serve as structural components of bones.

Minerals are also the major **electrolytes** in the circulatory system. Nutritionists divide minerals into two broad classes: the trace minerals and the major minerals. It is important to note that the terms trace and major do not reflect the importance of the mineral in the body, but rather the abundance of the mineral in the human body. For example, iron is considered to be a trace mineral, but it is crucial to the development of hemoglobin in the blood.

The digestive system handles minerals in a variety of ways. Some minerals, such as sodium and potassium, are quickly absorbed from food and transported by the circulatory system. However, some minerals, such as calcium, are poorly absorbed by the gastrointestinal (GI) tract. This level of potential availability of minerals is frequently called bioavailability and reflects not only the physical interaction of the digestive system with the mineral, but also the presence of certain chemicals in foods that may bind the mineral and make it unavailable to the digestive system. In addition, the ability of the GI tract to extract minerals from food is dependent on the overall health of the system, the age of the person, their sex, and other factors, such as pregnancy. The role of some of the more important minerals, such as iron, calcium, sodium, and potassium, will be discussed throughout this volume of the encyclopedia. By examining these minerals in detail, one can gain an appreciation of how the GI tract processes minerals in general.

Water

While the average person may not consider water to be a nutrient, in fact it is probably one of the most important nutrients for the digestive system. Like the circulatory, respiratory, and urinary systems, the digestive system is a water-based system that uses water to move nutrients, deliver digestive enzymes, lubricate the length of the gastrointestinal tract, and facilitate the absorption of nutrients into the circulatory and lymphatic systems. The average human requires about 2.65 quarts (approximately 2.5 liters) of water per day to meet the metabolic requirements of the body. The majority of this comes from liquids and foods that are consumed throughout the day. A smaller amount is derived from chemical reactions within the body, such as dehydration synthesis reactions.

The movement of water between the digestive system and the tissues of the body, most commonly the circulatory system, is highly regulated. The digestive system must simultaneously retain enough water for its own operation and supply the body with the water it needs to function. This is a complex task and frequently involves the use of minerals such as potassium and sodium to establish concentration gradients to efficiently move water. The large intestine, or colon, is the major digestive organ responsible for this process.

The Upper Gastrointestinal Tract: Oral Cavity, Esophagus, and Stomach

The human digestive system is actually a series of organs that form a long, enclosed tube. This organ system of the human body is specialized for breaking down incoming food into the needed nutrients for the body's vast array of metabolic functions. The majority of the organs in the human body are either directly or indirectly associated with the process of digestion. The organs of the GI tract are those that physically comprise the tube, also called the alimentary canal, which the food physically passes through. These include the oral cavity, esophagus, stomach, small intestine, and large intestine (also called the colon). Figure 3.1 gives the location of these organs in the human body. Associated with the organs of the gastrointestinal tract are the accessory organs. The accessory organs, which will be covered in detail later in this chapter, contribute needed materials for the breakdown and processing of the food entering the system. In many cases, the accessory organs have multiple functions that are highlighted in other volumes of this series.

For convenience, the GI tract is frequently divided into two major sections for study. The upper GI tract consists of the oral cavity, esophagus,

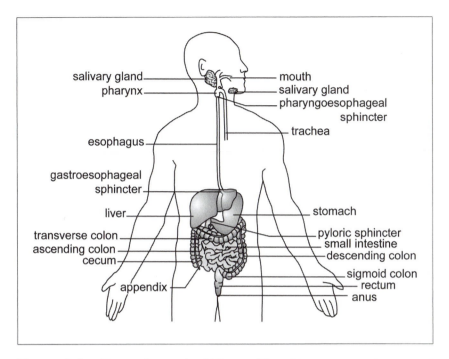

Figure 3.1 Gastrointestinal Tract. This diagram gives the general arrangement of the digestive and accessory organs. (Sandy Windelspecht/ Ricochet Productions)

and stomach, as well as associated valves and accessory organs. The lower GI tract consists primarily of the small intestine and colon. This division, while practical from the standpoint of a reference book, also has some basis in physiological function. As this chapter will explore, the role of the upper GI tract is primarily in the processing of food material. The majority of the digestion and nutrient processing, as well as the preparation of waste material, occurs in the lower GI tract.

The Oral Cavity

The human mouth, also called the oral cavity or **buccal cavity**, is the entry point into the human digestive system. The oral cavity represents an area of intense activity for the body. Not only are nutrients initially processed in this location, but is also serves as the connecting point between the respiratory system and the outside environment, as well as the location of a significant amount of sensory input from chemical receptors, most notably taste (Figure 3.2). The human mouth is the site of both mechanical and enzymatic digestive processes.

Salivary Glands

Vital to the digestive functions of the oral cavity are the secretions of three pairs of accessory glands collectively called the salivary glands. These

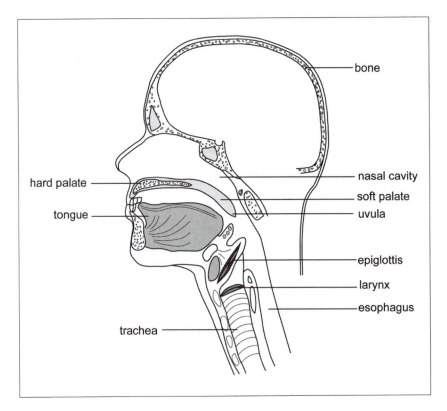

Figure 3.2 Oral Cavity. The structures of the oral cavity, including the relationship to the nasal cavity and respiratory passages. (Sandy Windelspecht/Ricochet Productions)

glands are identified by their location in the oral cavity. Two pairs are located along the bottom of the oral cavity. The sublingular glands are located just below the tongue, and the submandibular glands are positioned just beneath these, near the mandibula (jawbone). A third set, called the parotid glands, are located just in front of, and slightly below, the ears. The childhood disease called mumps frequently infects the parotid glands, although the other pairs may become infected as well. A duct carries the secretions of each salivary gland into the oral cavity.

Saliva, the chemical secretion of the salivary glands, is actually a complex mixture that performs a variety of functions for the digestive system. Saliva is primarily water (99.5 percent), which serves to lubricate and moisten the digestive system. However, it is the remaining 0.5 percent of the volume that contains some of saliva's most important functions. This small fraction contains important **ions**, such as potassium, chloride, sodium, and phosphates, which serve as pH buffers and activators of enzymatic activity.

Since salivary glands are similar in structure to sweat glands found within the skin, they also secrete urea and uric acid as waste products. Saliva also contains a small amount of an enzyme called lysozyme, which inhibits, but does not eliminate, the formation of bacterial colonies in the oral cavity. Mucus, a watery mixture of complex polysaccharides, helps lubricate and protect the oral cavity. Also found in saliva is another enzyme, called salivary amylase, which initiates the process of carbohydrate digestion (discussed at length in the following section).

The composition of the saliva varies slightly depending on the salivary gland in which it originates. The salivary glands of an adult can secrete a combined volume of 1.58 quarts (1,500 milliliters) of saliva daily. The amount of saliva secreted at a specific time is dependent on a number of factors. For example, when the body is dehydrated, the production of saliva is decreased, which in turn contributes to a thirst response by the body. An increase in saliva production is under direct control of the brain and is usually the result of a response to a chemical stimulus. The sight or smell of food typically serves to increase saliva production. Memories can also result in an increase of saliva production, such as the memory of a favorite food or food-related event. Increased saliva production can continue for some time after eating to cleanse the mouth of food, eliminate harmful bacteria, and restore the normal pH of the oral cavity.

Mechanical Digestion

In the oral cavity, the process of mechanical digestion serves several functions. First, the action of the teeth and tongue break the food into small
portions so that it may be sent to the stomach via the esophagus. Second,
the process of mechanical digestion increases the surface area of the food,
allowing the secretions of the salivary glands to mix freely with the food
and stimulating the action of the taste buds.

The action of chewing, or mastication, is the first stage of mechanical
digestion. Chewing involves the action of both the teeth and the tongue.
While there are three major types of teeth in a human adult mouth (molars
and premolars are frequently classified as one type), all teeth have the same
fundamental structure (Figure 3.3). It is the shape of the tooth that determines its function in mechanical digestion (Table 3.2). The combination
of different types of teeth in the mouth allows for the processing of a large
variety of foods, from protein-rich meats to nutritious vegetables and fruits.
The shape and structure of the human jaw is designed to provide a large
physical force to the teeth, which can be used to grind both plant tissue
and bones to release nutrients.

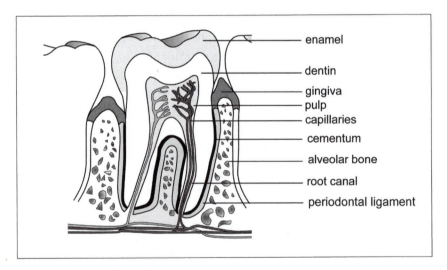

Figure 3.3   The Teeth. The internal structure of a typical human tooth.
(Sandy Windelspecht/Ricochet Productions)

TABLE 3.2
The Teeth of an Adult Human

Class	Function	Number
Incisors	Cutting food	8
Cuspids (canines)	Tearing and shredding	4
Premolars	Crushing and grinding food	8
Molars	Crushing and grinding food	12

The teeth absorb the brunt of this force. To prevent damage, each tooth is located in a socket of the jawbone. Connecting each tooth to the socket is the periodontal ligament, which also acts as a shock absorber. The socket and lower portions of the tooth are covered by the gums, or gingivae. (A common inflammation of this tissue is called gingivitis.) Teeth are made from a calcified form of connective tissue called **dentin**, which is covered with a combination of calcium phosphate and calcium carbonate commonly called enamel. (Dental caries typically erode this area of the tooth.) Within the center of each tooth is an area called the pulp cavity, which contains nerves, blood vessels, and ducts of the lymphatic system. Without teeth, humans would be required to swallow food whole in much the same manner as snakes. As organisms with a high metabolic rate, we require a relatively rapid processing of incoming nutrients. The importance of the teeth in increasing the surface area of the food for later enzymatic digestion should not be underestimated.

The second stage in the mechanical processing of the food involves the action of the tongue. The tongue is comprised of skeletal muscle, which is under the voluntary (but not always conscious) control of the body. The movement of the tongue is controlled by two separate sets of muscles. The extrinsic muscles enable the movement of the tongue that is important for digesting food. These muscles move the food from the area of the teeth to the back of the mouth, where it is formed into a small round mass of material called a **bolus**. This area at the rear of the oral cavity is commonly called the **pharynx**. The pharynx serves as the junction between the respiratory system and digestive system, and thus all activity in this area must be highly coordinated by the body. By the action of the tongue, the food is lubricated with saliva to facilitate swallowing and to mix in the enzymes of the salivary glands. The tongue also participates in the swallow reflex (see the "Swallowing Reflex" section) through the action of the intrinsic muscles. This muscle group also controls the size and shape of the tongue and is involved with speech.

Located on the tongue are a series of **papillae**, which are small projections of the tissue. It is the papillae that give the tongue its rough texture. The papillae are sometimes mistakenly referred to as the taste buds, but the taste buds are actually specialized receptors located at the base of certain types of papillae. There are three different forms of papillae, which differ in their appearance and location on the tongue. The circumvallate papillae are the largest and are located in a V-shaped region at the rear of the tongue (Figure 3.4). All of the circumvallate papillae contain taste buds. The fungiform papillae are knoblike in appearance and are dispersed across the entire tongue. Depending on their location, some fungiform papillae contain taste buds. The last group is the filiform papillae. These have a

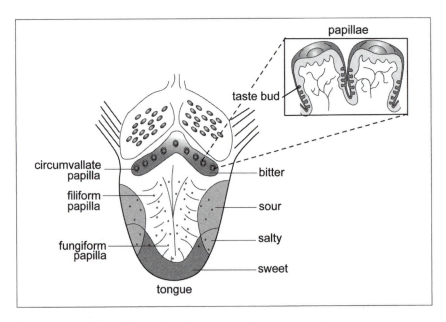

Figure 3.4   The Tongue. The external structure of the tongue showing the relationship between papillae types and taste zones. (Sandy Windelspecht/Ricochet Productions)

filament-like appearance and are also uniformly distributed across the surface of the tongue. However, unlike fungiform papillae, filiform papillae rarely contain taste buds.

The tongue is divided into four different taste zones—sour, bitter, sweet, and salty. The taste buds in each area are sensitive to a unique chemical signature. Depending on the origin of the signal, the brain interprets the different tastes. The number of receptors that fire and the duration of the signal determine the intensity of the taste.

Enzymatic Digestion

Enzymatic digestion is responsible for breaking organic material into smaller subunits that can be absorbed into the circulatory system. The amount of enzymatic digestion within the oral cavity is small in comparison to the activity of the lower GI tract. However, there is some initial digestion of both carbohydrates and lipids in the oral cavity. The salivary glands, primarily the submandibular and sublingual glands, secrete an enzyme called salivary amylase. Recall that the nutrients are primarily absorbed from the digestive system in their simplest structure, or monomers. Salivary amylase belongs to a class of enzymes that digest complex carbohydrates, such as starch, into monosaccharides. The monosaccharides are easily absorbed into the circulatory system, although little absorption occurs in the oral cavity. The salivary amylase is mixed into the food by the action of the tongue and cheeks and continues to break down the starches in the food for about an hour until deactivated by the acidic pH of the stomach. A second enzyme of the oral cavity is lingual lipase. Lingual lipase is secreted from glands on the surface of the tongue. This enzyme acts on triglycerides in the food, breaking them down into monoglycerides and fatty acids. However, the action of this enzyme is relatively minor and it does not make a major contribution to overall lipid digestion.

Swallowing Reflex

As the bolus forms in the rear of the oral cavity, or pharynx, the swallowing reflex begins. Swallowing, or **deglutition**, is a staged process that is partly under voluntary control and partly a reflex action. While most people do not consciously think of swallowing, in fact it represents a complex, highly coordinated activity. The tongue, through the action of the intrinsic muscles, forces the food to the back of the mouth. The pressure of the bolus on the pharynx activates a series of receptors that send a signal to the swallowing center of the **medulla oblongata** and **pons** in the brain. The swallowing center then temporarily deactivates the respiratory centers of the brain to ensure that the bolus will be directed into the digestive, and not the respiratory, system.

As the bolus prepares to enter the esophagus, a series of events is initiated to direct the food into the digestive system. Once the swallowing reflex has begun, the following four events occur in rapid succession:

1. The tongue moves upward against the roof (hard palate) of the mouth to prevent the food from reentering the oral cavity.

2. The uvula, an inverted-Y-shaped flap of skin at the rear of the mouth, moves upward to block the nasal passages.

3. The vocal cords in the larynx tightly close over the opening of the windpipe, or glottis.

4. As the bolus passes into the esophagus, it forces a flap of cartilaginous tissue called the epiglottis downward over the glottis as an added precaution to protect against the food entering the respiratory system.

Layers of the Digestive System

Before following the bolus on its brief journey through the esophagus, it is necessary to discuss the tissue structure of the digestive tract. From the esophagus to the anus, the walls of the digestive tract have the same general structure, with minor variations in each organ to enable specific functions. Within the wall of the digestive tract are four major tissue layers. From outermost to innermost, they are the serosa, muscularis externa, submucosa, and muscosa.

The serosa is the outermost layer of the digestive tract and is comprised of connective tissue. The serosa is important in that it forms a connection between the digestive tract and the **mesentery** that suspends the organs of the digestive tract within the abdominal cavity. To prevent friction between the organs of the system, the serosa secretes a water-based mixture that lubricates the exteriors of the organs. Directly under the serosa is a double layer of smooth muscle, the muscularis externa. These two muscle layers are the inner circular muscle and the outer longitudinal muscle. Since these layers are composed of smooth muscle, they are not under voluntary control of the brain. However, a nerve network called the myenteric plexus allows for regulation of activity from the involuntary control centers of the brain. The muscles contract in different directions, with the circular layer controlling the diameter of the digestive tract and the longitudinal layer controlling the length. The human digestive system does not rely on gravity to move nutrients through it; instead, the action of these two muscle layers rhythmically moves food through the system by a series of coordinated contractions called **peristaltic action**.

The next layer inward is a dense section of connective tissue called the submucosa. Located within the submucosa are the major blood and lymphatic vessels, as well as another series of nerves called the submucous plexus that provides involuntary regulation of the layer. The innermost layer is the mucosa. This layer lines the interior of the digestive tract and thus is in direct contact with the nutrients passing through the system. The **epithelial cells** of the mucosa serve several functions, depending on the region of the gastrointestinal tract. In some cases, these cells secrete a mucus layer that serves to lubricate the passage and protect the cells. Other cells may secrete digestive juices, while still others may release hormones that regulate the activity of the region. Epithelial cells are arranged into folds to increase the surface area. The amount of folding is dependent on the region of the gastrointestinal tract. Also within the mucosa, usually just underneath the epithelial cells, is a thin layer of smooth muscle, as well as blood vessels, lymphatic vessels, nerves, and cells of the immune system.

Esophagus

Once the bolus leaves the oral cavity, it enters into a muscular tube called the esophagus. The esophagus is not a major digestive organ, because the only enzymes that are active here are the salivary amylase and lingual lipase from the oral cavity. Furthermore, since the bolus spends only a brief amount of time in the esophagus (between five and nine seconds), and since the mucosa tissue layer does not contain a large number of folds, there is almost no absorption of nutrients through the walls of the esophagus. Instead, the esophagus serves as a conduit from the oral cavity, through the thoracic region of the body and to the stomach. The thoracic region houses the heart and lungs and is bordered on the bottom by a muscular barrier called the **diaphragm**. The esophagus passes through the diaphragm and connects to the upper portion of the stomach.

The bolus moves through the esophagus by peristaltic action. To aid the movement of the bolus, the cells of the mucosa tissue layer secrete mucus to lubricate the tube. To ensure the one-way movement of food, the esophagus is regulated by two sphincters, or valves. At the upper end of the esophagus is the pharyngoesophageal sphincter, which also serves to limit the flow of air into the gastrointestinal tract during breathing. At the lower end of the esophagus is the gastroesophageal sphincter, sometimes called the cardiac sphincter, which connects the esophagus to the stomach. The gastroesophageal sphincter also inhibits the reflux, or backup, of gastric juices from the stomach into the esophagus. Without this valve, the highly acidic gastric juices would damage the delicate mucosa layer of the esophagus.

Stomach

The stomach is commonly recognized as a muscular sac that functions as a holding site for food before it enters into the small intestine, as well as the location where the food is mixed and partially digested by mechanical processes. However, while the stomach does perform these functions, in actuality, its physiology and role in the digestive process is much more complex.

The stomach is an elastic, J-shaped organ whose boundaries are defined at the upper end by the gastroesophageal sphincter, and at the lower end by the pyloric sphincter (Figure 3.5). When empty, a human stomach may have a volume as little as 0.05 quarts (50 millimeters). In comparison, when full, the stomach may contain almost 1.06–2.11 quarts (1–2 liters) of food, depending on the individual. The stomach contains the same tissue layers as are found in the esophagus, small intestine, and colon, with some important variations in the secretions and structure of the mucosa layer. The J-shaped interior of the stomach is divided into regions based on slight differences in the secretions of the mucosa layer, thickness of the muscle layers,

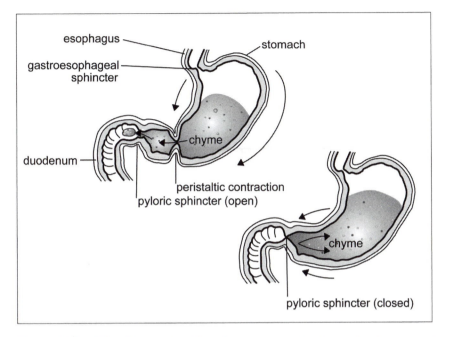

Figure 3.5   The Stomach. This diagram not only shows the location of the sphincters that define the boundaries of the stomach, but also the peristaltic contractions that are responsible for moving forward toward the duodenum. (Sandy Windelspecht/Ricochet Productions)

and overall function in digestion. The uppermost part, located above the level of the gastroesophageal sphincter, is the fundus. Below this is the main region of the stomach, called the body. The lower portion of the stomach, which connects to the small intestine, is called the antrum. A fourth region, called the cardia, is located around the area of the gastroesophageal opening and plays only a limited role in digestion. The physiological differences of the fundus, body, and antrum will be covered within the following sections.

Composition of Gastric Juice

The stomach produces about 2.12 quarts (2 liters) of gastric juice per day. Recall that in the general structure of the digestive tract, the mucosa tissue layer may contain folds. In the stomach, these folds are called gastric pits (Figure 3.6). Located in each of these pits are specialized cells that are

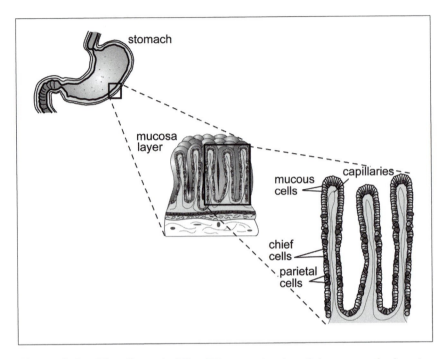

Figure 3.6 The Gastric Pits. The gastric pits of the stomach showing the location of chief and parietal cells. (Sandy Windelspecht/Ricochet Productions)

responsible for generating the secretions of the stomach. At the top of the gastric pit are the mucous neck cells. Together with the surface epithelial cells, they are responsible for secreting the mucus coating that protects not only the cells within the gastric pit, but also the mucosa layer of the stomach. The mucus has an **alkaline**, or basic, pH, and thus serves to neutralize any stomach acid before it comes in contact with the stomach mucosa. The mucus also serves to lubricate the interior of the stomach for mechanical digestion. Located within the gastric pits are the chief cells and parietal cells. The chief cells secrete an inactive enzyme called pepsinogen, which is involved in the chemical digestion of proteins (as described in the next section).

The parietal cells are responsible for manufacturing hydrochloric acid. The manufacture of hydrochloric acid is an energy-intense process, and thus the parietal cells have an exceptionally high concentration of mitochondria to generate the needed energy. Hydrochloric acid has a pH of 2.0, making it 100,000 times more acidic than water. The hydrochloric acid serves a number of functions in the stomach. First, it distorts, or denatures, the structure of proteins, making them easier to digest. Second, the low pH of the hydrochloric acid activates the pepsinogen enzyme secreted by the chief cells. Finally, the low pH of the hydrochloric acid acts as a deterrent against bacterial contaminants in the food. In this regard, the gastric juice of the stomach acts as a physical barrier of the immune system. In addition to hydrochloric acid, the parietal cells secrete an intrinsic factor that aids in the absorption of vitamin B_{12} in the small intestine.

While the gastric pits of the fundus, body, and antrum may look fundamentally the same, there are some important variations in the secretions from these areas. Gastric juice, containing hydrochloric acid, pepsinogen, some mucous, and intrinsic factors, is secreted primarily by the cells of the fundus and body. In comparison, the cells of the antrum are responsible for secreting a large amount of mucus. Specialized cells in this area, called G cells, release a hormone called gastrin. Gastrin is released directly into the bloodstream and regulates the activity of the parietal and chief cells in the body and the stomach.

Enzymatic Digestion

The food was mixed in the oral cavity with saliva, which contains salivary amylase. The chemical digestion of complex carbohydrates in the bolus continues down the esophagus. After the bolus passes through the gastroesophageal sphincter and enters the stomach, it comes in contact with the highly acidic hydrochloric acid. While hydrochloric acid deactivates salivary amylase, the lack of a significant amount of mechanical digestion in the fundus and upper regions of the body of the stomach allows the salivary amylase to continue carbohydrate digestion within the bolus. However, once mechanical digestion begins lower in the stomach, the salivary amylase is quickly inactivated. The thick mucus coating of the stomach prohibits the absorption of digested carbohydrates into the bloodstream.

In the oral cavity, the enzyme lingual lipase initiated a limited digestion of triglycerides in the food. In the stomach, the chief cells also release a gastric lipase, which serves much the same function in breaking down triglycerides.

As was the case with the carbohydrates in the oral cavity and stomach, this is not a major contribution to the overall digestion of these nutrients, and there is no appreciable absorption of triglycerides through the stomach lining.

The prime nutrient target of enzymatic digestion in the stomach is protein. Recall from earlier in this chapter that proteins may be large molecules, and all contain multiple levels of complex organization. This three-dimensional structure of proteins, and the presence of peptide bonds holding the amino acids together, makes proteins a difficult class of nutrients to digest. The purpose of protein digestion in the stomach is to initialize the process by destabilizing the structure of the protein. Thus, the mechanisms of protein digestion in the stomach are very general, and not directed at one specific type of protein. As was the case with carbohydrates, there is no absorption of peptides or amino acids through the lining of the stomach.

Enzymes that are involved in protein digestion belong to the general class called proteases. The pepsinogen secreted by the chief cells in the gastric pits is initially inactive, so as to protect the cells of the gastric pit from unintentional digestion. After being secreted, the pepsinogen makes its way through the protective mucus coat and into the main cavity, or **lumen**, of the stomach (Figure 3.7). The hydrochloric acid in the lumen activates the pepsinogen by cleaving off a small fragment from one end of the molecule. This active form of the enzyme is called pepsin. Pepsin also has the ability to activate pepsinogen in what is frequently called an **autocatalytic process**. Once activated, pepsin breaks down some proteins into smaller peptide fragments for further digestion later in the small intestine.

Mechanical Digestion

The stomach is primarily an organ of mechanical digestion, whose purpose is to thoroughly mix the incoming food material with gastric juice, forming a semi-solid mixture called chyme. This process occurs in three distinct stages: (1) the filling of the stomach with food and the temporary storage of food, (2) the mixing of the food with gastric juice, and (3) the emptying of the stomach. The purpose of these processes is to manipulate the chyme to the correct consistency, so that it can pass through the pyloric sphincter into the upper region of the small intestine, also called the duodenum, for digestion. A series of complex signals between the stomach and small intestine controls

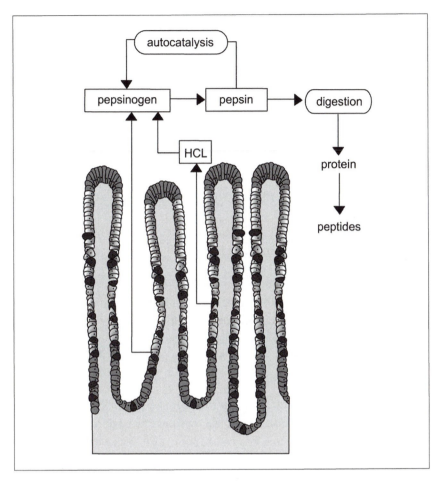

Figure 3.7   Digestive Enzymes. The autocatalytic activation of digestive enzymes in the stomach. (Sandy Windelspecht/Ricochet Productions)

the final movement of materials into the duodenum. As was the case with enzymatic digestion, each of the three regions of the stomach has slightly different roles in mechanical digestion. Certain surgical procedures can also alter a body's digestion patterns and are used as a means to treat severe obesity (Sidebar 3.1).

After passing through the gastroesophageal sphincter, the bolus enters into the body of the stomach. With a normal volume of about 1.5 fluid

Treating Severe Obesity through Surgery

Diet and exercise are important approaches to treating overweight and obesity. However, in the case of severe obesity, diet and exercise are not possible and are not enough. Another treatment option is bariatric surgery, which assists weight loss and has been shown to reduce chronic health conditions associated with obesity—such as type 2 diabetes—through restricting food intake and blocking the digestive system's absorption of calories and nutrients. Of course, healthy eating and exercise must follow the bariatric surgery for optimum results.

There are three primary bariatric surgeries commonly offered in the United States. Below are some details on each of these procedures:

Adjustable Gastric Band: Also known as AGB, this surgery works primarily by reducing food intake by placing a small band around the top of the stomach. This produces a small pouch that is about the size of a thumb. The size of the pouch is controlled by a balloon located inside the band. It can be inflated or deflated with saline solution, based on the needs of the patient.

Roux-en-Y Gastric Bypass: This surgery—RYGB—works by decreasing food intake, in addition to decreasing food absorption. Similar to the AGB surgery, a pouch is created about the same size to the one using the adjustable band. Absorption of food is reduced by directing food directly from the pouch to the small intestine, bypassing the stomach and upper intestine.

Biliopancreatic Diversion with a Duodenal Switch: Also known as BPD-DS or the duodenal switch, this surgery has three parts. The procedure removes a large part of the stomach to reduce portion sizes, routes food away from the small intestine to diminish absorption, and re-routes bile and other digestive juices, which impair digestion. When a large portion of the stomach is removed, a gastric sleeve in a tubular shape is created. This is known as a vertical sleeve gastrectomy, or VSG. This sleeve is connected to a short segment of a part of the digestive system known as the duodenum, which is directly connected to a lower part of the small intestine. The smaller portion of the duodenum remains available for food intake, as well as vitamin and mineral absorption. But in fact, the majority of food bypasses duodenum. The surgery reduces the distance between the stomach and colon, which encourages malaborption. The result can be significant weight loss, although there are risks of long-term complications due to the reduced absorption of vitamins and minerals.

---

ounces (50 milliliters), this region of the stomach would quickly fill with food if it were not for the elastic nature of the stomach lining. Located along the inside of the stomach are deep folds of tissue called rugae. The purpose of the rugae is to allow the gradual expansion of the stomach, eventually allowing a liter or more of food to enter into the cavity. This process is called receptive relaxation, since it involves the gradual relaxation of the rugae to accommodate the incoming food. This allows the stomach to easily expand its volume to about 1.06 quarts (1 liter), after which the tension of the stomach may cause discomfort. The volume at which this occurs varies with the individual, dietary habits, emotional state, and a number of other factors.

The peristaltic contraction of the smooth muscle that is responsible for mixing the food with gastric juice to produce chyme is initiated in the fundus. However, the areas of the stomach differ in the strength of the smooth muscle and thus the intensity of the peristaltic action. The contraction in the fundus is relatively weak, but becomes progressively stronger as it moves through the body and antrum. In the body of the stomach the contractions are not sufficiently powerful enough to provide a significant amount of mixing, allowing the continued digestion of carbohydrates in the bolus by salivary amylase. Thus in many regards the body of the stomach acts primarily as a storage site for incoming food.

As the contractions continue, the gastric juice mixes with the food to form chyme. This chyme is propelled downward into the narrower regions of the antrum, where the force of the peristaltic actions increases significantly. At the terminal end of the antrum is the pyloric sphincter, which serves to isolate the stomach from the small intestine. The pyloric sphincter is never completely closed, allowing for an almost continuous passage of water and other fluids into the duodenum. As the peristaltic contraction of the antrum approaches the phyloric sphincter, a small amount of chyme is moved through the opening and into the duodenum. However, the majority of the chyme is blocked from passing and is forced backed into the antrum for further mixing.

Regulating Stomach Motility

The emptying of the stomach contents, also called motility, usually takes between two to four hours following completion of a meal and is dependent on a large number of factors. These factors either inhibit or stimulate the movement of the chyme and are summarized in Table 3.3. For the most part, actions of the stomach increase motility into the duodenum, while feedback from the duodenum inhibits movement of chyme through the pyloric sphincter. There are three distinct phases to stomach motility: the cephalic phase, gastric phase, and intestinal phase.

The prefix ceph- means "head," and the cephalic phase refers to the interaction of the brain with the stomach. If chemical receptors detect the smell or taste of food, a signal is sent to the medulla oblongata in the brainstem, which relays a signal along the vagus nerves to the submucosal plexus in the stomach. The submucosal plexus then stimulates the activity of chief and parietal cells, thus preparing the stomach for incoming food. A similar event occurs when a person thinks about food, especially those foods that that the individual enjoys. The emotional state of the individual, such as anger or anxiety, may inhibit these stimuli by activating the **sympathetic nervous system**. The sympathetic nervous system is involved with the

TABLE 3.3
Factors Influencing the Movement of Chyme into the Duodenum

Stimulation	Inhibition
Distention of the stomach	Distention of the duodenum
Presence of partially digested proteins in stomach	Presence of fatty acids and carbohydrates in the duodenum
Gastrin	Cholecystokinin (CCK) Gastric inhibitory peptide (GIP)
Fluid chyme	Viscous chyme
Presence of alcohol or caffeine in stomach	

fight-or-flight response, one aspect of which is to reduce activity in the gastrointestinal system so that blood may be redirected to muscles.

As its name implies, the gastric phase involves the activity of the stomach. Two factors influence its activity. First, the amount of distention, or stretching of the stomach lining, acts as an indicator of the fullness of the stomach. As the stomach fills, and the rugae relax, stretch receptors in the lining stimulate the release of gastrin by G cells in the mucosal lining of the antrum. Gastrin is released into the blood stream, where it returns to the stomach to stimulate the generation of gastric juice by the parietal and chief cells. The gastric juice has a normal pH of around 2.0; if this becomes more basic (or alkaline), then the secretion of gastrin is increased. (A reverse reaction occurs if the pH level of the stomach increases above 2.0.) The stretch receptors also stimulate the peristaltic contractions of the stomach.

As noted in the section on mechanical digestion, these contractions are responsible for mixing the incoming food with gastric juice. As the contractions increase in strength, more of the chyme passes through the pyloric sphincter. The amount varies with the consistency, or fluidity, of the chyme. Under the ideal conditions, around 0.01–0.02 quarts (10–15 milliliters) of material may pass through the pyloric sphincter with each wave of contractions.

The duodenum of the small intestine may also regulate the activity of the stomach during the intestinal phase. Since the small intestine represents the major organ of digestion and absorption in the body, the duodenum must be ready to receive the incoming chyme for processing. The duodenum primarily has an inhibitory effect on stomach motility (see Table 3.3). Distention of the duodenum, due to the presence of a large volume of chyme, initiates a neural response called the enterogastric reflex, which through the action of the medulla oblongata decreases the strength of peristaltic contractions in the stomach, thus reducing the amount of chyme entering the duodenum.

The presence of partially digested carbohydrates and fats in the duodenum also activates an inhibitory pathway, but this pathway is based on the action of hormones. The action of salivary amylase, lingual lipase, and gastric lipase had previously started digestion of both carbohydrates and fats in the stomach. When these breakdown products reach the duodenum, they signal the release of gastric inhibitory peptide (GIP), secretin, and cholecystokinin

(CCK) by the mucosa layer of the duodenum. These hormones are released directly into the bloodstream and influence the activity of a number of organs of the digestive tract, including the stomach. The secretions of the stomach are inhibited by secretin and GIP, while CCK and GIP reduce gastric motility. When this occurs, fewer breakdown products are generated, less chyme enters the duodenum, and thus fewer hormones are produced. When the gastric and intestinal regulatory mechanisms are combined, it allows a fine-tuning of the gastrointestinal system to ensure that the optimal amount of material is being processed by the small intestine at all times.

Absorption of Nutrients

As mentioned previously, very few nutrients are absorbed through the lining of the stomach, primarily due to the presence of the mucus layer, which isolates the mucosa tissue from the hydrochloric acid. However, water and some ions are able to be absorbed directly into the circulatory system. In addition, both ethyl alcohol (the form found in alcoholic beverages) and acetylsalicylic acid (commonly known as aspirin) are able to penetrate the mucus layer and enter into the circulatory system.

The Lower Gastrointestinal Tract: Small Intestine and Large Intestine (Colon)

The previous section described how the upper gastrointestinal (GI) tract, consisting of the stomach, esophagus, and oral cavity, was involved with the processing of food for digestion. While there were some examples of enzymatic digestion in the upper gastrointestinal tract, the majority of the activity was associated with mechanical processing. The primary purpose of this mechanical digestion was to increase the surface area of the food so that the enzymes of the lower gastrointestinal tract can efficiently break down the nutrients into chemical forms that are able to be rapidly absorbed by the small intestine.

The lower gastrointestinal tract consists of two digestive organs called the small intestine and large intestine. Assisting with the operation of the small intestine are three accessory organs, the liver, gall bladder, and pancreas (Figure 3.8). The lower gastrointestinal tract serves two primary functions. First, the small intestine functions as the main organ of digestion

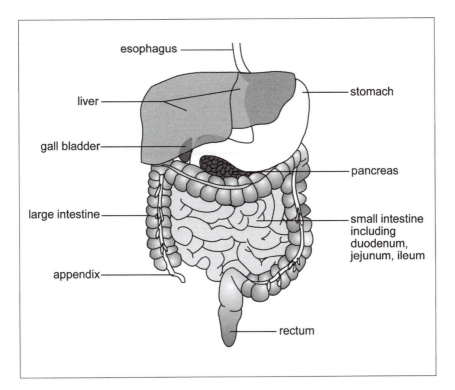

Figure 3.8 The Lower Gastrointestinal Tract. The lower gastrointestinal tract, giving the locations of the digestive organs and accessory glands. The large intestine is frequently called the colon. (Sandy Windelspecht/Ricochet Productions)

and absorption in the human body. It is here that the bulk of nutrient processing is performed. The large intestine, or colon as it is commonly called, is primarily involved in the reabsorption of water and salts back into the body, and the preparation of the fecal material for excretion. This section covers the physiology of the small and large intestine and their interaction with the accessory organs of the digestive system.

Small Intestine

The name of the small intestine is derived from its diameter, and not its overall size. The small intestine averages only approximately 1 inch (2.5 centimeters) in diameter, but in an adult can be over 10 feet (3 meters) in length. Although it may be of a small diameter, the small intestine represents the major site of digestion and absorption in the human body and is thus one of the more important organs of the digestive system.

The small intestine connects to the stomach at the pyloric sphincter and empties into the colon through the ileocecal valve, also called the ileocecal sphincter. For the purpose of study the small intestine is divided into three segments, although there is little difference in the physical appearance or structure of the regions. The first 7.8–9.8 inches (20–25 centimeters) of the small intestine, starting at the pyloric sphincter, is the duodenum. The next 2.7 yards (2.5 meters) of the small intestine is called the jejunum, and the last section, about 2 yards (2 meters) in length and ending at the ileocecal valve, is the ileum.

This chapter previously examined the structure of the tissue layers in the gastrointestinal tract. As was the case with the stomach, the physical characteristics of these tissue layers vary in the lower gastrointestinal tract. The most significant of these differences occur in the mucosa and submucosa layers, the two innermost tissue layers. These layers interact directly with the interior cavity, or lumen, of the small intestine.

The most notable difference in the structure of the mucosa layer is the presence of numerous fingerlike projections called villi. Unlike the folds in the mucosa layer of the stomach, which enable it to expand in response to an incoming volume of food, the villi of the small intestine are involved in increasing the surface area to facilitate the absorption of nutrients. There may be as many as 40 villi per square millimeter of mucosa, effectively increasing the surface area of the small intestine by a factor of 10. Within each villi are capillaries and portions of the lymphatic section called **lacteals** (Figure 3.9). As nutrients are absorbed into the villi (see the following section), they pass into either the capillaries or lacteals and are transported away from the small intestine. Along each of the villi are located four types of specialized cells. At the base of each villi, in pits called the intestinal glands (also known as the crypt of Lieberkühn), are the Paneth cells. These cells release lysozyme, an enzyme that protects the small intestine from bacteria. They also may move larger nutrient particles out of the lumen by the process of **phagocytosis**. Also located within the intestinal glands are enteroendocrine cells. As their name implies, these cells are actually part of the endocrine system and are

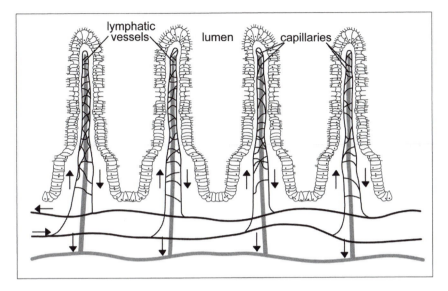

Figure 3.9   Villi of Small Intestine. Notice that each villi contains both capillaries and lymphatic vessels. (Sandy Windelspecht/Ricochet Productions)

responsible for releasing hormones such as gastric inhibitory peptide (GIP), secretin, and cholecystokin (CCK).

Further up the villi are located the goblet cells, which are responsible for secreting the protective mucus coating of the small intestine. While this mucus coating is not as thick as that found in the stomach, it serves to lubricate and protect the mucosa of the small intestine. Located along the length of the villi, but primarily above the region of the intestinal glands, is a layer of epithelial cells. These cells, also called absorptive cells, are the major site of nutrient absorption in the small intestine. These cells are unique in that the plasma membrane on the lumen side of the cell contains a large number of small projections called microvilli. Each of these projections is about 1 micrometer (μm) long and a typical absorptive cell may have as many as 6,000 microvilli on its surface. A square millimeter of small intestine may contain up to 200 million microvilli. This increases the surface area of the small intestine by an additional factor of 20. When combined, the villi and microvilli of the small intestine increase the overall surface area in this organ by 600 times, further disproving the idea that this is a "small" intestine.

Under a microscope, the microvilli appear as a thin, fuzzy barrier on the lumen side of the absorptive cells. This is sometimes called the brush border, and it represents a region that is not only involved in nutrient absorption, but also in nutrient digestion by a group of enzymes called the brush border enzymes (see the following sections on the individual nutrient classes).

In the submucosa, a group of specialized cells called the duodenal glands, also known as the Brunner's glands, release additional mucus into the lumen. This mucus is alkaline in pH, which helps to neutralize any hydrochloric acid from the stomach remaining in the food as it moves through the small intestine.

Movement of Nutrients

It takes approximately three to five hours for nutrients to transit the small intestine from the pyloric sphincter to the ileocecal valve. In the stomach, the incoming food was mixed with gastric juice to form chyme. This chyme is moved along the length of the small intestine by two different types of contractions, peristalsis and segmentation. The peristaltic contractions in the small intestine are similar to those found in the stomach and esophagus. However, the contractions in the small intestine are much lower in intensity than those in the upper gastrointestinal organs.

The primary mechanism by which the chyme is moved through the small intestine is by segmentation. Unlike peristalsis, which is the rhythmic, sequential contraction of the smooth muscle layers of the gastrointestinal tract, the process of segmentation involves the localized contraction of small segments of the small intestine. These circular contractions squeeze the chyme against the mucosa layer of intestine, bringing the nutrients into direct contact with the microvilli of the absorptive cells. These contractions also serve to further mix the chyme with the secretions of the small intestine, gall bladder, and pancreas (see the following section on the role of the accessory glands). As the chyme is mixed by segmentation, it is slowly propelled along the length of the small intestine toward the ileocecal valve. Since the contractions of segmentation are not directional, as is the case with peristaltic contractions, some of the material will actually back up into the previous segment of the intestine. To ensure that the chyme has an overall movement toward the large intestine, the duodenum contracts more frequently (around 12 per minute) than either the jejunum or ileum (approximately nine per minute).

After the processing of a meal is complete, the small intestine enters into a "housekeeping" mode to remove the remnants of the chyme from the lumen. This consists of a series of weak peristaltic contractions that begin in the duodenum and contract for a short length of the intestine before ending weakly. The next contraction begins a little further down the intestine, and so on. The process is analogous to a sweeping action and is called the migrating motility complex. The entire process can take several hours to complete.

Although the small and large intestine are linked by similarities in their names, in reality the internal environments and physiology of these organs are vastly different. As will be described later in this chapter, the large intestine possesses a natural population of bacteria. If allowed into the small intestine, these bacteria could wreak havoc with the delicate tissues present there. The ileocecal valve is well designed to prevent the movement of materials into the small intestine from the large intestine. The folds of the valve are arranged to open easily as the chyme moving through the ileum of the small intestine exerts pressure against the valve. However, if material in the large intestine presses against the valve, the pressure forces the folds tightly closed, preventing contamination of the small intestine. The valve is also under hormonal control. As food enters the stomach, special cells release a hormone called gastrin. Gastrin serves to relax the ileocecal valve, allowing for the emptying of the small intestine in response to an incoming meal.

Role of Accessory Glands

The digestive functions of the small intestine are assisted by the secretions of three accessory glands—the liver, gall bladder, and pancreas. This section will introduce the digestive roles of these glands.

One of the most important organs in the human body is the liver. The largest organ by weight, the liver coordinates activity between a number of body systems. The liver not only provides chemicals to assist in the digestive process, but it also filters and stores nutrients coming from the digestive system. As an accessory organ to the small intestine, the liver provides a compound called bile, which assists in the process of lipid digestion. Bile is synthesized in the small intestine from a number of chemicals, including cholesterol, phospholipids, bile acids, and water. Another ingredient is **bilirubin**, a waste product from the breakdown of worn-out red blood cells in the liver. Bilirubin does not play a role in digestion, but does give bile and the fecal material their color. Bile is continuously excreted into the bile duct, which connects the liver to the duodenum via the hepatopancreatic ampulla.

The gall bladder is a small sac, roughly pear-shaped, that is located just beneath the liver. The sole purpose of the gall bladder is the storage of the bile salts from the liver between meals. Before entering the duodenum, the bile duct links up with the pancreatic duct to form the hepatopancreatic ampulla. At the duodenum end of this structure is a small valve called the Sphincter of Oddi. This sphincter is normally open when chyme is present in the duodenum, but closes between meals. Since bile production by the liver is a continuous event, the bile leaving the liver backs up and enters the gall bladder to be stored. As food enters the duodenum, CCK is released from the enteroendocrine cells in the mucosa of the small intestine. This hormone signals the gall bladder to contract, releasing its contents, as well as acting as a signal for the Sphincter of Oddi to relax.

The third major accessory organ is the pancreas. It is located just below the stomach, adjacent to the small intestine. Like the liver, the pancreas performs a variety of functions in the human body. As an accessory organ for the digestive system, the pancreas is responsible for providing the majority of the digestive enzymes needed for nutrient processing in the small intestine. The cells of the pancreas produce a colorless liquid called pancreatic juice. This colorless, watery mixture is collected into the pancreatic duct, which later joins the bile duct to form the hepatopancreatic ampulla, which empties into the duodenum. The major enzymes present in the pancreatic juice are listed in Table 3.4. In addition to these enzymes, pancreatic juice contains a compound called sodium biocarbonate. This compound is slightly alkaline and serves to neutralize the acid in the chyme from the stomach. This establishes the correct pH for optimal enzyme activity in the small intestine (Sidebar 3.2).

Overview of Nutrient Processing

As mentioned previously, the small intestine is the primary organ of digestion and absorption in the human body. In general, the small intestine has

TABLE 3.4
The Sources of the Digestive Compounds of the Small Intestine

Target nutrient	Compound	Source
Carbohydrates	Pancreatic amylase	Pancreas
Lipids	Bile	Liver
	Pancreatic lipase	Pancreas
Proteins	Chymotrypsin	Pancreas
	Trypsin	Pancreas
	Carboxypeptidase	Pancreas
	Elastase	Pancreas
Nucleic acids	Deoxyribonuclease	Pancreas
	Ribonuclease	Pancreas

to process two very broad classes of nutrients. The **hydrophilic** ("water-loving") molecules, which include the monosaccharides and many of the amino acids, are easily transported, digested, and absorbed by the small intestine. The second class is the hydrophobic ("water-fearing") molecules, of which fats and cholesterol are the major examples. Due to their chemical properties, hydrophobic molecules will require more elaborate processing and transportation systems.

For each of the four major classes of organic nutrients, the carbohydrates, lipids, proteins, and nucleic acids, the job of the small intestine is to first break down the large complex structures of these nutrients into units that are small enough to be transported into the villi. The following sections will detail the digestion and absorption of each of the major classes of nutrients. It is important to remember that, for the most part, the processing of nutrients occurs simultaneously throughout the length of the small intestine.

Carbohydrate Digestion and Absorption
Carbohydrate digestion began in the oral cavity with the activity of the salivary amylase enzyme. While this enzyme only acts on the incoming food for a brief period of time before becoming inactivated by the hydrochloric acid of the stomach, it does initiate the breakdown of starches and other polysaccharides into disaccharides and monosaccharides. The majority of

SIDEBAR 3.2

Alcohol's Negative Impact on the Liver

The disease of alcohol has many negative health impacts, but it is especially harsh on the liver. One of the digestive system's secondary organs, the liver breaks down alcohol so it can be eliminated from the body. When the body consumes excessive amounts of alcohol, the liver has a hard time processing it fast enough. This leads to an imbalance—too much alcohol in the liver—which interferes with how the organ breaks down protein, fat, and carbohydrates. This can lead to alcohol-induced liver disease.

There are three primary types of alcohol-induced liver disease:

Fatty Liver Disease: This condition occurs when there is an accumulation of fat cells in the liver. The liver may be enlarged, and discomfort may be felt by patients in the upper abdomen.

Alcoholic Hepatitis: This occurs in up to 35 percent of heavy drinkers, and is characterized by inflammation of the liver. Patients may experience nausea, vomiting, fever, and jaundice. While this disease can cause progressive liver damage over a period of years, it can be reversed if drinking is stopped. Severe forms of alcoholic hepatitis can lead to life-threatening complications.

Alcoholic Cirrhosis: This is the most serious type of alcohol-induced liver disease, and occurs when normal liver tissue is replaced with scar tissue. It is the most serious type of alcohol-induced liver disease. An estimated 10–20 percent of heavy drinkers develop this condition, typically after drinking for more than 10 years. Symptoms are similar to alcoholic hepatitis. Cirrhosis is a life-threatening disease, and cannot be reversed, although damage can be minimized upon abstaining from alcohol.

Complications of these and other alcohol-induced liver diseases include a buildup of abdominal fluid, an enlarged spleen, bleeding from esophageal veins, high blood pressure in the liver, kidney failure, and liver cancer.

the polysaccharide digestion is conducted in the small intestine by a secretion of the pancreas called pancreatic amylase. Pancreatic amylase is active in the lumen of the small intestine. Like salivary amylase, this enzyme can break down starch and glycogen, but not plant polysaccharides such as cellulose, commonly called fiber. These remain undigested and unprocessed until they reach the large intestine and provide much of the bulk of the food.

To be absorbed into the villi, carbohydrates must be in their simplest form, as monosaccharides. In most cases, the action of the salivary and pancreatic amylases generates a compound called dextrin, a short chain of glucose molecules. The final breakdown of the carbohydrates into monosaccharides is performed by a group of enzymes physically embedded in the brush border of the small intestine. These enzymes are named by the specific carbohydrate substrate that they recognize. For example, the dextrinase enzyme breaks down the short-chains of glucose (dextrins) into single glucose units. The maltase enzyme digests the disaccharide sugar maltose into two glucose units. Sucrose, another of the disaccharides that consists of glucose and fructose, is digested by the sucrase enzyme, and lactose, a sugar commonly found in milk, is broken into the monosaccharides galactose and glucose by the action of the lactase enzyme.

Once the carbohydrates in the chyme have been digested into one of the three monosaccharides (fructose, glucose, or galactose), they are ready for absorption into the villi of the small intestine (Figure 3.10). The molecules themselves are too large to pass directly through the membranes of the intestinal cells. Fructose is absorbed by a process called **facilitated diffusion**, in which a protein channel in the membrane of the epithelial cells aids in the movement of the fructose into the villi. The movement of glucose and galactose is slightly different. To move these molecules into the epithelial cell, the cell couples their movement with the transport of sodium ions (Na^+) into the cell. This is a form of active transport, and as such requires energy.

Once these monosaccharides are in the cytoplasm of the epithelial cell, they are rapidly moved into the capillaries of the villi by the process of facilitated diffusion. There they enter the circulatory system and proceed to the liver for additional processing. The rapid movement of these molecules out of the epithelial cells ensures that the interior of the cell has a low concentration of the monosaccharides, thus aiding in the absorption of sugars from the lumen.

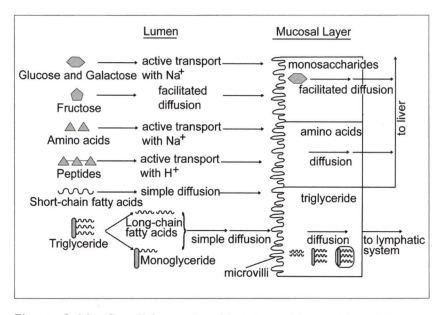

Figure 3.10   Small Intestine Nutrient Absorption. Nutrient absorption in the small intestine. Note how monosaccharides and most amino acids proceed directly to the liver, while the triglycerides enter in the lymphatic system. Triglycerides are first disassembled in the lumen and then reassembled into chylomicrons for transport in the lymphatic system. (Sandy Windelspecht/Ricochet Productions)

Proteins

The digestion of proteins was initiated in the stomach with the activity of the enzyme pepsin. Recall that pepsin is a general enzyme, which serves to disrupt the complex three-dimensional structure of the proteins in the chyme. Thus, when the proteins reach the small intestine, they are primarily in the form of small chains called peptides. In general, protease enzymes are powerful catalysts and thus are secreted in an inactive form until needed by the body. This prevents the unwanted digestion of the mucosal lining. For example, in the stomach, the pepsin enzyme was generated from a molecule called pepsinogen under the correct pH conditions. A similar event occurs in the small intestine. The pancreas releases a protoenzyme called trypsinogen into the lumen of the duodenum via the pancreatic duct. Trypsinogen is inactive and cannot begin the breakdown of proteins until it is first activated by enterokinase, an enzyme that is present in the brush border region of the villi. The enterokinase cleaves off a small portion of the trypsinogen molecule, forming the enzyme trypsin. In turn, trypsin activates three reactions with chymotrypsinogen, proelastase, and procarboxypeptidase to produce chymotrypsin, elastase, and caboxypeptidase, respectively. These molecules are all proteases that are active in the lumen of the small intestine. While each targets a different structural part of the protein, their overall function is to digest the proteins into smaller peptides, at times releasing individual amino acids.

Present in the brush border region are two other enzymes. The first are the aminopeptidases, which are responsible for removing the terminal amino acid from the end of the peptide chain. The second is an enzyme called dipeptidase, which breaks the remaining peptide bond holding two amino acids together. The end result of this activity is free amino acids, which may now be absorbed into the villi.

The absorption of the individual amino acids is very similar to the process outlined previously for carbohydrates. In most cases, the movement of amino acids across the membrane is an active process that is coupled to the movement of sodium ions (Na^+), although in some cases the process is coupled to the transport of hydrogen ions (H^+). Once in the epithelial cells, the amino acids diffuse into the capillaries of the villi, to be transported by the **hepatic portal system** to the liver for processing.

There are two additional aspects of protein digestion to be mentioned. While we previously noted that the three regions of the small intestine are fundamentally the same, there are some minor differences with regards to protein digestion and absorption. The majority of protein digestion occurs in the duodenum and jejunum, with only minimal activity in the ileum. Also, any protein that enters the small intestine is subject to the digestive process outlined above. This includes not only proteins from food sources, but also the proteins found in worn-out mucosal cells, enzymes such as pepsin and salivary amylase, bacterial proteins, and miscellaneous proteins such as bilirubin that are excreted from the liver and pancreas.

Lipids
The processing of lipids by the small intestine differs significantly than that described previously for carbohydrates and proteins. This is primarily due to

the fact that lipids are hydrophobic molecules, and as such are not easy to work within the hydrophilic environment of the lumen of the small intestine.

Prior to the entry of lipids into the small intestine, there was a small amount of processing performed by the lingual lipase in the oral cavity. This enzyme primarily serves to initiate triglyceride digestion, but is active only for a relatively short period of time before entering the stomach. In the stomach, gastric lipase may act on short-chain fatty acids, such as those found in milk products. However, this enzyme does not make a significant contribution to lipid processing. Thus, the real first level of lipid digestion and absorption occurs in the small intestine.

As the chyme passes the pyloric sphincter, it is mixed with secretions of the liver, gall bladder, and pancreas. Recall from the previous section that the liver produces bile salts, which may be temporarily stored and concentrated within the gall bladder. Bile emulsifies, or breaks down, droplets of fats in the chyme into smaller particles. This is not enzymatic digestion, since the individual lipid molecules are not the target, but rather the interaction between the fat molecules. The result is small droplets of lipids with a diameter of about 0.039 inches (1 millimeter). This drastically increases the surface area for digestion, speeding the overall processing of the lipids. The enzyme responsible for the digestion of lipids is pancreatic lipase. Pancreatic lipase enters the small intestine through the duct called the hepatopancreatic ampulla, along with bile from the gall bladder and liver. This lipase breaks down triglycerides into small fatty acid chains and monoglycerides, which consist of a single fatty acid chain connected to the glycerol backbone. In these forms, the molecules can then be absorbed directly into the epithelial cells of the villi by the process of diffusion.

However, most of the triglycerides that enter into the small intestine contain long-chain fatty acids, which due to their size cannot diffuse into the villi. For the long-chain fatty acids, and similar hydrophobic molecules such as cholesterol, a different process exists to move the nutrients out of the lumen and into the body. When combined with the bile salts released from the liver, the lipids and cholesterol form spheres called micelles. Each micelle consists of an outer shell of approximately 30 to 50 bile salt molecules. Micelles are **amphipathic molecules**, meaning that they have both polar and nonpolar regions, enabling them to interact with both hydrophobic and hydrophilic molecules. The hydrophobic lipids are carried within the center of the micelle. When the micelle reaches the cell membrane of the epithelial cells in the villi, the lipids and other hydrophobic molecules in the core of the micelle are able to diffuse across the membrane. The sphere of bile salts is then able to return to the lumen to pick up more hydrophobic lipids. In other words, the micelle acts as a shuttle by providing a hospitable environment for the movement of large hydrophobic molecules.

Since bile salts represent a reusable resource for the digestive system, they are recycled in the small intestine. The bile salts that were initially released in the duodenum are reabsorbed in the ileum of the small intestine. There they enter into the portal circulatory system and are returned to the liver. This circular recycling of bile salts is sometimes called the enterohepatic circulation.

One more important aspect of lipid processing occurs in the small intestine. With hydrophobic nutrients, such as sugars, the nutrients that are absorbed by the small intestine quickly diffuse into the capillaries of the villi, where they then enter the circulatory system. However, lipids by nature do not interact well with an aqueous environment, and their large size would quickly clog the narrow capillaries contained within the villi. Instead, lipids are packaged into a special form of lipoprotein called a chylomicron. Chylomicrons are protein-covered balls of lipids, cholesterol, and phospholipids. The role of the chylomicron is to move the lipids into the lacteal of the villi, where it then enters into the lymphatic system.

Nucleic Acids

Recall that nucleic acids represent the genetic material of living organisms, and thus are present in most of the material being processed by the small intestine. Since DNA and RNA (ribonucleic acid) both consist of long chains of nucleotides, they are digested in a similar manner by the small intestine. The fact that the DNA is double-stranded, and typically a longer polymer, has little influence on the properties of nucleic acid digestion.

Along with its previously mentioned enzymes, the pancreas secretes pancreatic ribonuclease and deoxyribonuclease, which act on RNA and DNA, respectively. The purpose of these enzymes is to cleave individual nucleotides from the polymer. In many ways these enzymes function similarly to the pancreatic proteases mentioned earlier. Once a nucleotide is removed from the polymer, it is further digested by brush border enzymes called nucleosidases and phosphatases. These enzymes break the nucleotide down into its constituent sugars, phosphates, and nitrogenous bases for absorption into the villi by active transport. Once there, they move into the capillaries by diffusion and enter the portal circulatory system.

Water

While water is an integral part of the digestive tract, and one of the more important nutrients in the human body, it is not "digested" in the same manner as the organic nutrients just mentioned. Instead, it is absorbed by the villi of the small intestine into the circulatory system, and to a lesser extent the lymphatic system. Most physiologists believe that water moves from the lumen of the intestine into the epithelial cells by the process of osmosis, or the diffusion of water. This passive process is dependent on the concentration of solutes and has long been recognized as the prime mechanism of water movement by biological systems.

However, researchers have begun to discover that many organisms possess specialized channel proteins, called aquaporins, that allow for the rapid movement of water across a plasma membrane. The small intestine processes a tremendous volume of water daily. Almost 10 quarts (close to 9.3 liters) of water enter the small intestine daily, most of it (7.4 quarts, or approximately 7.0 liters) comes from the secretions of the accessory glands (4.2 quarts, or about 4 liters), stomach (2.1 quarts; 2 liters), and small intestine (1.06 quarts; 1 liter). The remainder (2.4 quarts; an average of 2.3 liters) is obtained from the ingested food and liquids. The small intestine reabsorbs almost 90 percent of this volume, with the remainder passing into the large intestine.

Vitamins

Vitamins are frequently assigned to two general classes, those that are water soluble (vitamin C and the B-complex vitamins) and those that are fat-soluble (vitamins A, D, E, and K). These general classes also apply to the approach that the small intestine takes in absorbing these important compounds. The water-soluble vitamins are treated in much the same manner as monosaccharides and amino acids, meaning that they are actively transported into the epithelial cells and then move by diffusion into the capillary of the villi.

Fat-soluble vitamins may either move into the epithelial cells by diffusion, or through the action of the micelles. They are typically then loaded into chylomicrons for transport into the lymphatic system. Some vitamins have special processing in the small intestine. Vitamin B_{12}, sometimes also called cobalamin, is typically found in protein-rich foods. The pH of the stomach releases the vitamin, which then binds with an intrinsic factor before entering the small intestine. In the small intestine, vitamin B_{12} is absorbed into the epithelial cells, where it then returns to the liver via the enterohepatic circulation. The liver continuously secretes both vitamin B_{12} and folate into the bile. Since folate is associated with the health of rapidly dividing cells, and vitamin B_{12} is needed to activate folate, this mechanism ensures that the rapidly dividing epithelial cells of the small intestine are provided with a source of these important vitamins.

Minerals
Minerals, like water, are not organic nutrients and thus are not digested by the small intestine. However, the small intestine does represent an important location of absorption for many of the minerals in a human diet.

Unlike the other nutrient classes, mineral absorption in the small intestine is not guaranteed. Many foods contain chemicals that actively bind nutrients, reducing their ability to be absorbed. Thus, for minerals, it is often more correct to refer to their **bioavailability**, and not necessarily the total quantity in the food. Examples of these binders are oxalic acid, found in leafy vegetables such as spinach, and phylic acid, a compound frequently found in grains and beans (legumes).

There are many different minerals, each with its own unique absorption properties. For nutritional purposes, minerals are classified as being either trace or major, depending on the quantity that is required in the diet. However, since minerals are not digested enzymatically, as was the case with the organic nutrients, the activity of the small intestine is confined to absorption only. Most of the minerals behave as hydrophilic molecules (such as potassium and sodium), but a few display hydrophobic characteristics. In most cases, minerals are absorbed by active transport into the villi.

Two minerals of special interest in examining the physiology of the small intestine are calcium and iron. Calcium is an important nutrient for muscle contraction, and has the secondary function of providing strength to bone (see Chapters 7 and 11 of this encyclopedia for more information). Calcium is usually brought into the digestive system in the form of a salt, and is kept in a soluble form by the acidic nature of the stomach. The efficiency of the small intestine in absorbing calcium is based upon a number of factors. In general, an adult human is able to absorb about 30 percent of the calcium found in food, although this value may vary depending on age, sex, gastrointestinal tract health, and emotional state. For example, young children frequently absorb up to 60 percent of ingested calcium, and the value in pregnant women can reach 50 percent. (Other factors that influence calcium absorption are listed in Table 3.5.) However, the greatest factor that influences the absorption of calcium is the presence of vitamin D. Vitamin D, a fat-soluble vitamin, actually functions as a hormone, in that it is manufactured by one organ of the body to influence the activity of a second organ. Vitamin D may also be found in some foods, such as milk products, where it is frequently added to enhance calcium absorption. Once activated, vitamin D stimulates the small intestine to produce a calcium-binding protein, which in turn facilitates the movement of calcium into the villi.

The absorption of iron is slightly more complex. As was the case with calcium, all of the iron that is ingested is not absorbed. In fact, as little as 10 percent of the available iron is absorbed in an adult male, and only 15 percent in an adult female (Table 3.6). Iron may exist in one of two ion forms: ferrous iron (Fe^{2+}) or ferric iron (Fe^{3+}). Of these, ferrous iron is more easily absorbed. The source of the iron also plays a role in iron absorption. Iron that is present in animal flesh, called heme iron, is more easily absorbed that iron that originates in plant material (non-heme iron). As was the case with calcium, several environmental factors contribute to the absorption of iron from the chyme. Recently it has been discovered that vitamin C, a water-soluble antioxidant vitamin, has the ability to keep iron in its ferrous form,

TABLE 3.5
Factors Influencing Calcium Absorption in the Small Intestine

Inhibitory factors	Enhancing factors
Presence of oxalates and phylates	Presence of growth hormones
High fiber diets	Presence of lactose in the chyme
High phosphorus intake	Equal concentrations of phosphorus vitamin D

TABLE 3.6

Factors Influencing Iron Absorption in the Small Intestine

Inhibitory factors	Enhancing factors
Presence of phylates	Citric acid and lactic acid
High fiber diets	Some sugars
Presence of phosphorus and calcium	Vitamin C
Food additives such as EDTA (ethylenediamene tetra acetate)	MFP factor

thus increasing its bioavailability. In addition, many meat products contain a substance called MFP factor that increases iron absorption. However, the iron processing by the small intestine can also be inhibited by the presence of a number of compounds. One additional interesting feature of iron physiology is the ability of the small intestine to act as a temporary storage site. Unlike most nutrients, which quickly move through the absorptive cells of the villi into the circulatory or lymphatic systems, the mucosal layer may actually store iron using a special protein called mucosal ferratin. This protein binds iron and releases it to mucosal transferrin when needed by the body. Transferrins are iron-transport proteins. Mucosal transferrin then transfers the iron to blood transferrin (sometimes just called transferrin) for movement into the body.

Large Intestine

The name large intestine is derived from its diameter (2.5 inches; 6.5 centimeters), not its length (1.37 yards; 1.5 meters). The large intestine begins at the ileocecal valve, which serves as the boundary between the small and large intestines. The terminal portion of the large intestine, and the entire gastrointestinal tract, is the anus. The large intestine is vastly shorter than the small intestine and differs significantly both in anatomy and function. The large intestine is often mistakenly considered the location in the body where waste material is generated. In reality, this organ is more of a recycling center and temporary storage location than a waste disposal site. The large intestine is comprised of three distinct regions: the cecum, the colon, and the rectum (see Figure 3.11). The colon is subdivided into four zones,

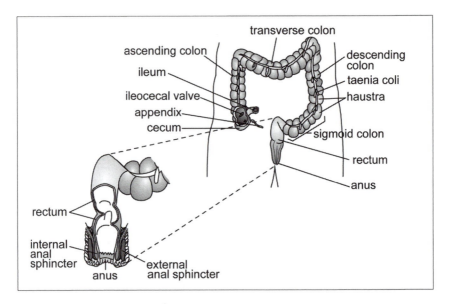

Figure 3.11   The Large Intestine. Commonly called the colon, the large intestine consists of three major sections, named for their orientation in the abdominal cavity. The large intestine terminates in the rectum. (Sandy Windelspecht/Ricochet Productions)

called the ascending colon, transverse colon, descending colon, and sigmoid colon, based on their orientation and position in the body cavity. The physiology of these zones is fundamentally the same, and the names are used for descriptive purposes only. Because the majority of the large intestine is comprised of the colon, the term colon is frequently used as a common name for the large intestine.

While the appendix has historically been considered a part of the large intestine, and often thought of as a vestigial organ, it is actually made from the same type of tissue as the lymph nodes and thus is now considered to be part of the lymphatic and immune systems. Lymphocytes housed in this area help protect the digestive system from pathogenic microorganisms, and thus the appendix serves as an important first line of defense for the lower GI tract. However, ailments of the appendix are frequently caused by problems with the digestive system.

Movement of Material

The large intestine receives approximately 15 fluid ounces (500 milliliters) of material daily from the small intestine. Under normal conditions, the majority of this material is the undigested remnants from the small intestine (see the following section on digestion and absorption). The colon regions of the large intestine are structured differently from the small intestine. Whereas the small intestine utilized circular and longitudinal patterns of smooth muscle to power the contractions necessary to move the food, the colon instead possesses three bands of smooth muscle. Rather than surrounding the GI tract, these muscles are arranged to run the length of the intestine, called the taeniae coli. The arrangement of the taeniae coli causes the exterior surface of the large intestine to resemble a series of small pouches, called haustra. The haustra are not permanent structures, but actually change position slightly based upon the contractions of the taeniae coli.

The movement of the material through the large intestine is a much slower process than in the small intestine. This gives ample time for the sections of the colon to reabsorb important nutrients such as water. In the small intestine, the contractions of the smooth muscle (segmentation) occurred at a rate upwards of 12 times a minute. In the large intestine, the contractions may occur several times an hour. These contractions are called haustral contractions, and the entire process is called haustration.

Also, unlike the small intestine, in which the contractions were controlled to move the chyme through the length of the intestine, in the colon, the haustral contractions are more regional. This causes the material to move back and forth between haustra, further increasing the time that the material is in the system.

Several conditions may cause a synchronization of these haustral contractions, resulting in a uniform movement of material toward the rectum and anus. The first of these is called a mass movement. Mass movements occur several times daily, usually following meals, and are characterized as synchronous contractions of the first two sections of the colon (ascending and traverse). This contraction propels the food into the descending colon, moving the material there into the sigmoid colon and rectum. Although a mass movement can occur without food entering the stomach, the body does possess a mechanism to clear the intestines to prepare for incoming food. As food enters the stomach, it triggers the gastrocolonic reflex. This reflex action causes contractions along the entire length of the intestines, moving food from the small intestine into the colon, and driving the undigested material in the colon into the rectum. The gastrocolonic reflex is often accompanied by an urge to defecate.

The final movement of material out of the gastrointestinal tract is called defecation and the factors that cause it to occur are called the defecation reflex. Defecation is actually a complex process since it involves both voluntary and involuntary actions of the anus. The anus, the terminal sphincter of the gastrointestinal tract, consists of both smooth and skeletal muscle. The internal anal sphincter is comprised of smooth muscle and thus is under involuntary control. During the defecation reflex, the smooth muscle of this sphincter relaxes. At the same time, the rectum and sigmoid colon contract, moving the contents toward the external anal sphincter. This sphincter is made of skeletal muscle, and thus is under voluntary control. If the external anal sphincter is relaxed, defecation occurs. If not, then the urge can be controlled, although this may result in excess water being absorbed from the feces, causing constipation. Although the external anal sphincter may be closed, it is still possible to force intestinal gas (flatus) out through a narrow opening in the anus, thus partially relieving pressure in the rectum.

Digestion and Absorption

Of the quarts of material that enter the digestive tract each day, only about 0.53 quarts (0.5 liters) actually ends up in the large intestine. The small intestine is highly effective as an organ of digestion and absorption; thus, the material reaching the large intestine usually contains only undigested material, such as cellulose, some water, salts, and bilirubin from the liver. Since the amount of usable nutrients is severely limited by the time the food material reaches the large intestine, there is no enzymatic digestion conducted by the cells of the large intestine. While no digestive enzymes are secreted by the large intestine, the mucosal cells along the length of the organ secrete an alkaline mucus, which serves to lubricate the internal lining of the large intestine and protect it from any acids produced by fermenting bacteria in this region. There is some minor breakdown of the bilirubin from the liver, and this accounts for the characteristic color of the fecal material.

The large intestine also lacks the complex internal structure found in the small intestine. Instead of a network of villi and microvilli, the interior surface of the large intestine is for the most part smooth. This reduced surface area limits the ability of the large intestine to be a major organ of absorption. However, the decreased surface area is slightly compensated for by slowing the movement of material through the organ. The relative lack of segmentation and peristaltic contractions in the large intestine means that the material is present in the intestine for a longer period of time, allowing for more (although slower) absorption of selected nutrients.

The action of the haustral contractions also serves to move the material back and forth within the colon, further slowing the movement of material. Dietary factors, namely the amount and types of fiber, also influence the rate of movement. In addition, health factors such as age, stress, and disease all contribute to the speed at which the material transits the large intestine. Of the 15 fluid ounces (500 milliliters) of the material that enters the large intestine, about 10.5 fluid ounces (350 milliliters) is reabsorbed, with the remaining volume exiting through the anus as feces.

The colon primarily absorbs water and salts, although it may also take in other nutrients, such as glucose and vitamins, that may be present (see the following section). Salts in the colon normally consist of both sodium and chloride ions, both of which are essential nutrients, and are reabsorbed. In addition, water is reabsorbed by osmosis, but a significant amount remains in the feces to lubricate it. The final daily fecal volume of 4.5 fluid ounces (150 milliliters) usually is two-thirds water and one-third solid material. Most of the solid material is actually bacterial mass, with bilirubin from the liver and cellulose accounting for the remainder.

The slower movement of material also gives microorganisms, such as bacteria, the opportunity to establish populations. However, unlike the remainder of the gastrointestinal tract, where the presence of bacteria causes problems, the colon actually contains a natural flora of bacteria that make a positive contribution to human physiology. The bacteria of the large intestine are in a symbiotic relationship with their human host.

The Accessory Organs

The previous two sections have examined the movement of food through the gastrointestinal system. As noted previously, there are two types of organs in the digestive system. Digestive organs, such as the stomach and small intestine, form the conduit through which food is moved and processed in the body. Associated with the operation of these organs are the accessory organs. In general, the accessory glands contribute important lubricants, enzymes, and chemicals that are required for the operation of the digestive organs. There are four accessory organs: the salivary glands, liver, gall bladder, and pancreas (see Table 3.7).

Salivary Glands

The salivary glands consist of three major pairs of glands that are located within the oral cavity. As previously mentioned, these are called the parotid, sublingual, and mandibular glands. There are also minor salivary glands, called the buccal glands, located in the linings of the cheek. Located under the tongue, along the base of the mouth, are the sublingual glands. Excretions from these glands are moved into the oral cavity by a short duct called the lesser sublingual duct. Located just below the sublingual glands, along the jawbone (mandibula), is the submandibular gland, which is connected to the oral cavity by the submandibular duct. Just in front of each

TABLE 3.7
Summary of Accessory Gland Contributions to Digestion

Organ	Role in digestion
Salivary glands	Provides salivary amylase for CHO digestion
	Provides mucus to lubricate oral cavity and esophagus
Liver	Manufactures bile for lipid digestion
	Central role in carbohydrate, fat, and protein metabolism
Gall bladder	Stores bile
Pancreas	Manufactures sodium bicarbonate
	Manufactures pancreatic digestive enzymes

ear are the parotid glands. These are connected to the oral cavity via the parotid duct, which enters close to the molars in the rear of the mouth.

The salivary glands play several important roles in the overall physiology of the oral cavity. First, they actively moisten the oral cavity, which greatly aids not only in the swallowing of food, but also in the process of speech. The mucus content of saliva serves to protect the tissues of the tongue and cheeks from the action of the teeth. In addition, saliva contains an antimicrobial compound called lysozyme, which serves to reduce, but not completely eliminate, bacterial growth in the mouth. Saliva is slightly alkaline in pH, and thus helps to buffer the oral cavity to the correct pH. The salivary glands also release an enzyme called salivary amylase, which initiates carbohydrate digestion. Each of the salivary glands varies slightly in the content of the saliva, although the saliva from each contains ions, mucus, water, and salivary amylase.

Combined, the salivary glands secrete an average of 1.06 quarts (1,000 milliliters) of saliva daily. The level of saliva production is dependent on a number of factors. In response to dehydration, the body limits saliva production, producing the thirst response. However, it is important to note that the feeling of a dry mouth lags the actual need for water, meaning that a dry mouth signals an advanced stage of dehydration. Most people are aware of increased saliva production in response to the sight or smell of food. This is due to the action of the nervous system, which has the ability to stimulate saliva production based on chemical signals from taste buds or olfactory (smell) glands, or by the touch of food on the tongue. It is also possible to invoke salivation by the memory of food, especially when hungry. The body may also increase or decrease saliva production during illness. In the case of fever or other illnesses, the body may reduce saliva production to conserve water (Sidebar 3.3). During times of nausea, saliva production may be increased.

Gall Bladder

The gall bladder is a small sac-like organ, 3.1–3.9 inches (8–10 centimeters) in length, located just under the liver (Figure 3.12). It is connected to the duodenum of the small intestine by the common bile duct. The gall bladder represents the simplest of the accessory glands in the fact that it primarily

SIDEBAR 3.3

Not to Be Underestimated: The Importance of Clean Water

Access to clean, safe water can not be undervalued when it comes to the maintaining the health of the digestive system on a global level. Proper sanitation and hygiene is vital to the health of societies everywhere in the world. According to the Centers for Disease Control and Prevention (CDC), eating contaminated food and drinking contaminated water can lead to developing certain infectious diseases related to certain germs, such as Cryptosporidium, Giardia, Shigella, and norovirus. Below are some important facts from the CDC about the importance of global access to safe water, adequate sanitation, and proper hygiene:

- Clean water as well as proper sanitation and hygiene have the potential to prevent at least 9.1 percent of the world's disease burden and 6.3 percent of all global deaths.

- Improved sanitation could save the lives of 1.5 million children every year who otherwise succumb to diarrhea-related diseases, according to the World Health Organization (WHO) and UNICEF.

- Over 800 million people around the world do not have access to an improved water source.

- An estimated 2.5 billion people—more than 30 percent of the world's population—do not have access to adequate sanitation.

- By improving water sources, deaths from diarrhea can be reduced by 21 percent worldwide. Improved sanitation can reduce deaths related to diarrhea by 37.5 percent; and simple hand washing at certain critical times can reduce the incidence of diarrhea by 35 percent. Improvement of drinking-water quality, including disinfection efforts, would lead to a 45 percent reduction of diarrhea cases.

- Millions of people around the world are infected with neglected tropical diseases (NTDs). NTDs are related hygiene-related, and are often found in locations with unsafe drinking water and poor sanitation. Examples of these diseases include Guinea Worm Disease (GWD), buruli ulcer, trachoma, and schistosomiasis. GWD, for example, is a rare but extremely painful parasitic infection spread through contaminated drinking water. Symptoms of GWD include thread-like worms slowly emerging from the human body through blisters.

- According to the United Nations and UNICEF, one in five girls worldwide who are of primary-school age are not in school, compared to one in six boys. One reason cited for this is lack of sanitation facilities for girls reaching puberty, or who are already going through menstruation. In addition, girls are also often responsible for collecting water for their family, making it difficult for them to attend school during school hours. For these reasons, installation of toilets and latrines may enable more schoolchildren, especially menstruating girls, to remain committed to getting an education.

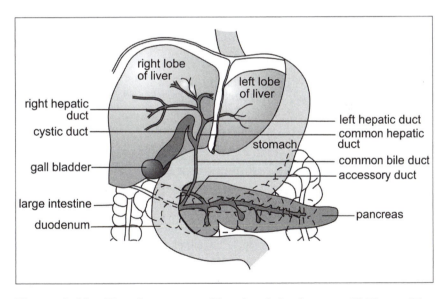

Figure 3.12   The Accessory Glands of the Lower GI Tract. The relationship of the accessory glands of the lower gastrointestinal tract to the stomach, small intestine, and large intestine (colon). (Sandy Windelspecht/Ricochet Productions)

serves as a storage location for secretions of the liver (see the next section). Unlike the pancreas, liver, and salivary glands, the gall bladder does not produce any chemicals necessary for digestion. Its sole purpose is the storage of bile between meals. Bile is produced by the liver and aids the small intestine in the digestion of hydrophobic molecules such as triglycerides.

The common bile duct is actually a conduit from the liver to the duodenum. At the junction of the small intestine is the sphincter of Oddi. When chyme is present in the duodenum, the sphincter of Oddi is open and the bile produced by the liver proceeds directly into the lumen of the small intestine. However, when chyme is absent, the sphincter is closed and the bile being continuously produced by the liver backs up in the bile duct and enters the gall bladder via a small duct called the cystic duct. Once stored in the gall bladder, the bile is concentrated and readied for the next meal. When the next meal enters the duodenum, the hormone CCK stimulates the gall bladder to contract, releasing the concentrated bile into the bile duct and into the duodenum.

Liver

The liver represents one of the most important and unique organs of the human body. From a genetic perspective, the cells of the liver are interesting in that some are **polyploid** and **binucleate**. Normally, cells of the body have a single nucleus and are **diploid**, meaning that they contain two copies of each chromosome. However, about 50 percent of the hepatocytes in the liver are polyploid cells, meaning that they contain additional copies of each chromosome, while others contain an extra nucleus. There are examples of hepatocytes that have eight or more copies of each chromosome. This arrangement most likely explains the large numbers of organelles found in these cells. Hepatocytes have some of the most abundant endoplasmic reticulum and Golgi bodies of any human cells, which enable them to manufacture large quantities of many biologically important molecules for export. The chromosomal and nuclear state of these cells may also explain regenerative properties of the liver.

With the exception of the skin, the liver is the largest organ in the human body. In an adult, the liver may weigh up to approximately 3 pounds (1.4 kilograms). The liver consists of two primary lobes, called the right and left lobes. The right lobe, the larger of the two, is sometimes subdivided for study into two additional lobes, called the quadrate and caudate lobes. The lobes are separated by a ligament called the falciform ligament. The falciform ligament not only defines the two major lobes, but together with other minor ligaments, it helps suspend the liver from the diaphragm. Despite the large size and fairly complex shape of the liver, its tissue is relatively homogenous, which plays an important role in liver physiology.

The liver is also a special organ in that it has the ability to regenerate itself in case of injury or disease. This is primarily due to the redundant structure of liver tissue. Each lobe of the liver consists of self-sufficient sub-units called lobules (Figure 3.13). Within each lobule are liver cells called hepatocytes and phagocytic cells called Kupffer's cells. The purpose of the Kupffer's cells is to engulf worn-out blood cells and invading pathogens, such as bacteria and viruses, arriving from the digestive tract. Thus, these cells technically belong to the immune, and not the digestive, system.

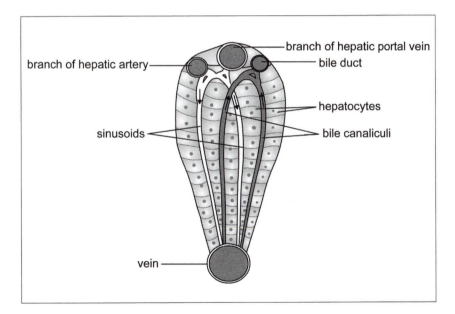

Figure 3.13   Lobule of the Liver. The liver is a redundant structure, meaning that each lobe of the liver contains numerous identical lobules. (Sandy Windelspecht/Ricochet Productions)

The hepatocytes are the cells where the work of the liver is conducted. On one side of each hepatocyte is a sinusoid. Blood from the digestive system, arriving via the portal circulatory system, enters into the sinusoid cavities. Sinusoids are not the same thing as capillaries, but rather represent open spaces from which the hepatocytes can extract nutrients from the digestive system. The phagocytic Kupffer's cells are located along the lining of the sinusoids to protect the liver from pathogens arriving from the digestive tract.

On the other side of the hepatocyte are small vessels called the bile canaliculi (bile canals). The hepatocytes continuously produce bile from cholesterol, lecithin (a phospholipid), and bile salts and secrete it into the bile canaliculi. Within each lobe, these vessels merge into larger structures called the left and right hepatic ducts, which in turn combine to form the common hepatic duct. The common hepatic duct carries bile to the gall bladder, where it then becomes the common bile duct. The common bile duct connects with the duodenum of the small intestine through the sphincter of Oddi.

Due to the liver's central role in digestive system physiology, there is a minor deviation in normal blood flow with regard to digestion. Typically, blood leaves the heart via arteries and proceeds to an organ of the body where it enters a capillary bed. The blood then returns to the heart by way of veins. However, in the processing of nutrients, there is a minor deviation in this path. Blood leaving the stomach, small intestine, and colon proceeds directly to the liver via the hepatic vein (Figure 3.14). This minor detour, sometimes called the portal, or hepatic, circulatory system, ensures that nutrient-rich blood from the digestive tract is first processed and screened by the liver, thus establishing the status of the liver as the master control organ for human digestion. Since the blood from the digestive system is low in oxygen, the oxygen needed for the metabolic functions of the liver cells is delivered by the hepatic artery.

The liver plays many important roles in human physiology. Since all of the blood leaving the digestive tract first passes through the sinusoids of the liver, the hepatocytes have the ability to screen and filter nutrients and other materials from the blood before it is delivered to the remainder of the body tissues.

In addition to screening, the liver secretes several important compounds. As previously noted, the hepatocytes manufacture bile salts from

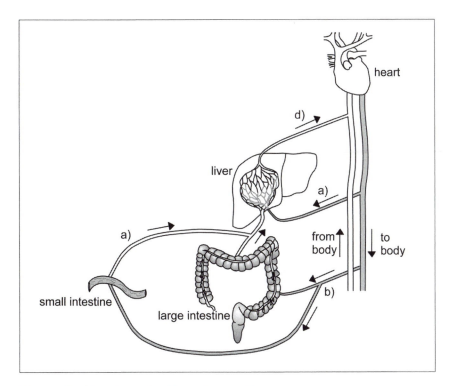

Figure 3.14 Hepatic Circulatory System. Blood leaving the heart may proceed directly to the liver (a) or to the intestines (b) by way of arteries. However, nutrient-rich blood from the intestines first returns to the liver (c) where nutrients are filtered before the blood returns to the heart (d). (Sandy Windelspecht/Ricochet Productions)

cholesterol. This bile is either secreted into the duodenum, or stored in the gall bladder. In addition, the liver excretes a compound called bilirubin, which is derived from the destruction of worn-out red blood cells. It is bilirubin that gives bile its characteristic yellow color. In the small intestine, bacteria break down bilirubin into stercobilin, giving the fecal material its brown color. However, a small amount is reabsorbed by the blood system and eventually excreted by the kidneys. This small amount of bilirubin is responsible for urine's yellow color.

Pancreas

The pancreas is an irregular-shaped gland that is located just below the stomach and adjacent to the duodenum of the small intestine. It averages between 4.7 and 5.8 inches (12 and 15 centimeters) in length, and a little over 0.8 inches (2 centimeters) in thickness. For descriptive purposes, it is divided into three major sections, although there is little difference in the physiology of the sections. The head is located closest to the duodenum and is connected to the digestive tract by two ducts. The hepatopancreatic duct is a common duct formed by the linking of the bile duct and pancreatic ducts. A second duct, called the duct of Santorini, directly connects the pancreas to the duodenum. Moving away from the duodenum and the head of the pancreas are the regions called the body and tail.

The pancreas actually represents two separate organs, both of which contribute to digestion, which are integrated into a single structure. A portion of the pancreas is an **exocrine gland**, meaning that it secretes compounds into a cavity. The second major area of the pancreas is the endocrine tissue, which secretes chemicals into the bloodstream. In general, the exocrine functions of the pancreas can be described as those directly involved with the processing of nutrients in the duodenum, while the endocrine is best described as those functions that involve hormones and the regulation of glucose **homeostasis** in the body. Both types of tissue exist throughout the pancreas.

Two types of cells make up the endocrine portions of the pancreas (Figure 3.15). Duct cells secrete what is formally called the aqueous alkaline solution. This solution is primarily sodium bicarbonate ($NaHCO_3$), and its purpose is to neutralize hydrochloric acid coming through the pyloric sphincter along with the chyme. These cells are named due to their close proximity to the pancreatic ducts. Deeper within the pancreas are groups of cells called acinar cells. These cells are responsible for generating the enzymatic secretions of the pancreas and together may excrete 1.6–2.12 quarts (1–2 liters) of fluid per day into the duodenum. The enzymatic secretions produced by the acinar cells of the pancreas contain three basic classes of enzymes. These are the proteolytic enzymes (proteins), pancreatic lipase (lipids), and pancreatic amylase (carbohydrates). It is these enzymes that enable the small intestine to conduct its physiological function as the major

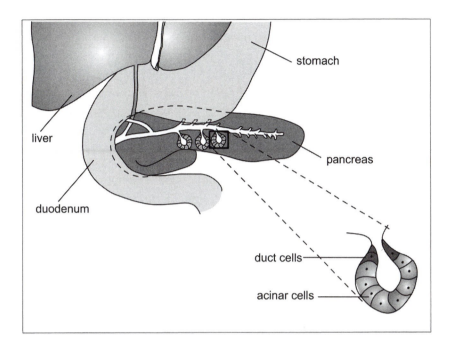

Figure 3.15 The Pancreas. Diagram of the internal exocrine structure of the pancreas showing the relationship between acinar and duct cells. (Sandy Windelspecht/Ricochet Productions)

organ of digestion and absorption. The pancreatic amylase and lipase enzymes hydrolyze carbohydrates and fats, respectively, into their monomers, which are then absorbed by the small intestine. However, the actions of the proteolytic secretions are a little more complex.

The proteolytic enzymes released by the pancreas are initially inactive. This protects the pancreatic cells from being damaged. The proteolytic excretions of the pancreas contain a mixture of enzymes: trypsinogen, chymotrypsinogen, proelastase, and procarboxypeptidase. Once released by the acinar cells, these inactive enzymes proceed through the pancreatic duct into the duodenum. Once in the lumen of the small intestine, rypsinogen is activated by an enterokinase enzyme located in the mucosal layer of the duodenum. The active form of trypsinogen is trypsin. Activated trypsin is an autocatalytic enzyme, meaning that it has the ability to react with additional trypsinogen to produce trypsin. Trypsin also activates chymotrypsinogen, proelastase and procarboxypeptidase to produce chymotrypsin, elastase, and carboxypeptidase, respectively. Together, these enzymes are responsible for digesting proteins in the food into individual amino acids or short chain peptides for absorption by the small intestine.

Summary

All living organisms, including human beings, take nutrients from the environment and process them into chemical compounds to fuel their cells. This is called metabolism, and is part of the larger process of digestion, which is when food is broken down into its essential elements and components. Although some compounds behave differently (such as water, minerals, and some vitamins), the body transports most of the digested food particles throughout the body. The digestive system is made up of primary and secondary organs. The primary organs are the mouth, esophagus, stomach, small intestine, and large intestine. The secondary or accessory organs are the liver, pancreas, salivary glands, and gall bladder. The primary organs communicate with other organs of the body, while the accessory organs enable the chemical interactions necessary for digestion.

Digestion is a vital function of the body, because it ensures there is an adequate source of energy available at all times. In addition, digestion provides the necessary materials to create and repair cells and tissues.

4

The Endocrine System

Stephanie Watson and Kelli Miller Stacy

Interesting Facts

- Ancient peoples called the pineal gland the "third eye" because they believed it held mystical powers. French philosopher René Descartes (1596–1650) thought that the pineal gland was the point at which the human soul met the physical body.

- The endocrine pancreas contains about one million small endocrine glands called the islets of Langerhans.

- Special fluid sensors in the brain, called osmoreceptors, are so sensitive that they can detect a 1 percent fluctuation in the body's water concentration.

- The adrenal cortex produces more than 60 different steroid hormones, but only a handful of these hormones are important to body function.

- Aldosterone, a steroid hormone produced in the adrenal cortex, acts upon the sweat glands to reduce the amount of sodium lost in the sweat. After a few days in a hot climate, sweat becomes virtually salt-free.

- During times of stress, the adrenal cortex can produce up to 10 times the normal amount of cortisol.

- The fetal adrenal gland is larger than the adult gland in relation to body mass.

- In recent years, scientists have discovered that many industrial chemicals, pesticides, and heavy metals interfere with the endocrine systems of humans and wildlife by mimicking natural hormones.

- Endocrine disruptors have even been found in the breast milk of Inuit women in the remote Arctic, where known endocrine-disrupting chemicals are neither used nor produced.

- Scientists say that some deli wrap, food can linings, teething rings, vinyl toys, medical IV bags, and plastic bottles may seep small amounts of potential endocrine disruptors.

Chapter Highlights

- Hormones and how they behave

- Target cells and receptors

- Second messenger systems

- Hormone regulation and secretion

- Feedback loops

- Glands of the endocrine system: hypothalamus; pituitary gland; anterior pituitary, posterior pituitary, pineal gland, thyroid, and parathyroid glands; adrenal glands; pancreas; ovaries; and testes

Words to Watch For

Adrenocorticotropic hormone	Antidiuretic hormone	Corticotroph
Agonists	Autocrine	Cortisol
Androgens	Calcitonin	Cytokines
Antagonists	Catecholamines	Dopamine
Anterior pituitary	Chondrocytes	Eicosanoids
	Corpus luteum	Electrolytes

Epinephrine

Erythropoietin

Estrogen

Exocrine glands

Extracellular fluid

Follicle-stimulating hormone

Gastric inhibitory peptide

Gastrin

Gene transcription

Glycogenolysis

Glycoprotein

Gonadotroph

Gonadotropins

Growth factors

Growth hormones

High-density lipoproteins

Homeostasis

Hypocalcemia

Hypophsis

Hypothalamic-hypophyseal portal system

Hypothalamic-pituitary-target organ axis

Hypothalamus

Inhibin

Insulin

Insulin-like growth factor

Intermediate pituitary

Intracellular fluid

Islets of Langerhans

Ketone bodies

Lactrotroph

Leptin

Low-density lipoproteins

Luteinizing hormone

Luteolysis

Mineralocorticoids

Motilin

Neurohormones

Neurosecretory cells

Norepinephrine

Oogenesis

Osmoreceptors

Pancreatic polypeptide

Paracrine

Parathyroid hormone

Polyunsaturated fatty acids

Posterior pituitary

Pregnenolone

Preprohormone

Progesterone

Progestins

Proglucagon

Proinsulin

Prolactin

Prostaglandins

Receptor

Seminiferous tubules

Sertoli cells

Somatostatin

Somatotroph

Substance P

Target cells

Testosterone

Thyroid-stimulating hormone

Thyrotroph

Thyrotropin-releasing hormone

Thyroxine

Triiodothyronine

Tyrosine

Vasopressin

Zona fasciculate

Zona glomerulosa

Zona reticularis

Introduction

For all of the various cells and tissues in the human body to work in harmony, they must communicate and coordinate with one another. Communication is essential in the human body. Without a network to integrate functions of the organs, muscles, nerves, and all other tissues, the body would virtually shut down. Ingested food would not be properly absorbed and utilized for energy, fluids and electrolytes would swing wildly up and down, and disease would easily set in, all with devastating effects.

To avoid these scenarios, the body has not one, but two integrated command centers. These centers—the nervous system (which is covered in Chapter 8 of this encyclopedia) and the endocrine system—act as the body's control towers, sending out messages that coordinate the function of every cell. The nervous system sends out its messages via electrical impulses, which travel within nerve cells (neurons); and chemical signals (called neurotransmitters), which transmit those impulses across small gaps (called synapses) between the neurons. The endocrine system sends out its messages via chemical messengers called hormones, which travel through the bloodstream to act on cells in other parts of the body. Together, the two systems regulate every essential function, from metabolism to growth and development.

The parts and functions of the endocrine and nervous systems are closely connected and synchronized. Nerves oversee the release and inhibition of endocrine system hormones, as well as blood flow to and from endocrine glands. Hormones, in turn, direct the nervous system by stimulating or inhibiting the release of neural impulses.

To comprehend the role of the endocrine system and the pathways by which hormones affect biologic processes, it is first necessary to understand the chemical makeup of hormones, to learn how they are secreted and by what mechanisms they reach and interact with their target tissues, and to discern the complex relationship between the endocrine and nervous systems. In this chapter, the anatomy of the glands will be discussed. These glands are the hypothalamus, pituitary, thyroid, parathyroids, adrenal, sex glands (testes and ovaries), and the endocrine pancreas. It is important to note that there are also tissues throughout the body that are not considered classic endocrine organs but that secrete hormones or hormone-like substances.

Fundamentals of the Endocrine System: Hormones and Their Actions

The process by which the endocrine system coordinates bodily functions is complex, consisting of many interrelated parts and systems that oversee hormone production, secretion, and delivery. At the core of the system are the hormones—the messengers that transport endocrine commands throughout the body.

Hormones are the chemical signals by which the endocrine system coordinates and regulates functions such as growth, development, metabolism, and reproduction. The word "hormone" comes from the Greek word meaning "to set in motion." When a hormone is released into the bloodstream, it does just that: It sets in motion a chain of events that ultimately results in a desired reaction within cells that are receptive to its influence. The reaction is generally designed to either trigger or inhibit a physiological activity.

Hormones are produced by various tissues and secreted into the blood or **extracellular fluid**. According to the traditional definition, hormones travel through the bloodstream to work on tissues in distant parts of the body. But some hormones act locally without ever entering the bloodstream. They may exert their effects on cells close to where they were produced (called **paracrine** action), or they may act on the same cells that produced them (**autocrine** action).

Some hormones act upon just one type of cell; others influence many different cells. Similarly, some cells are receptive to only one hormone, while others respond to several hormones. Hormones are not the only substances in the body that exert physiologic control on cells. A number of other chemical messengers act much like hormones. These include:

- *Neurotransmitters*: The nervous system, like the endocrine system, transmits messages to target tissues. But nervous system messages are made up of chemical signals called neurotransmitters.

Unlike endocrine cells, which release their hormones into circulation, neurotransmitters are released into the gap where two neurons (cells in the brain that receive and transmit nerve impulses) meet (called a synapse). At the synapse, they bind to receptors on the receiving neuron. Some substances (such as **epinephrine**, **norepinephrine**, **dopamine**, **gastrin**, and **somatostatin**) serve double duty, acting as both hormones and neurotransmitters.

- *Growth factors*: Not to be confused with **growth hormones** produced by the endocrine system, **growth factors** are proteins that bind to receptors on the cell surface and stimulate or inhibit cellular division and proliferation. Some growth factors act on many different kinds of cells; others target one specific cell type. Examples include platelet-derived growth factor (PGDF), epidermal growth factor (EGF), transforming growth factors (TGFs), and **erythropoietin**.

- *Cytokines*: Cytokines are signaling peptides secreted by immune cells (as well as by other types of cells) in response to stress, allergic reaction, infection, or other potentially harmful stimuli. Much like endocrine hormones, cytokines either travel through the bloodstream or act locally on target cells. Once cytokines bind to their receptors, they trigger a biological effect within cells. Cytokines may influence cell growth, cell activation, or cell death (i.e., in the case of cancer cells). They also act directly upon the **hypothalamic-pituitary-target organ axis** of the endocrine system by increasing or decreasing hormone synthesis as part of the body's stress response. There are four major categories of cytokines: interleukins, interferons, colony stimulating factors, and tumor necrosis factors (TNF).

- *Eicosanoids* (fatty acid derivatives): These compounds are produced from **polyunsaturated fatty acids**, most commonly from the precursor arachidonic acid. Depending upon which enzymes act on arachidonic acid, it may be converted into one of several classes of hormone-like substances, including **prostaglandins**, prostacyclines, and thromboxanes. Although they are not technically hormones, **eicosanoids** act in much the same way to influence a variety of physiological processes, including smooth muscle contraction; kidney, immune system, and reproductive function; and calcium mobilization. Eicosanoids primarily exert a paracrine influence on nearby cells or an autocrine influence on the cells that produced them.

Just as a hormone does not always have to fit the classic definition, a hormone-producing tissue does not always need to reside within the endocrine system. Although hormones are primarily associated with the endocrine glands, tissues in other parts of the body (kidneys, liver, and heart, for example) can also produce and release them.

Hormone Modes of Action

Hormones are grouped according to their chemical structure (see Table 4.1). The structure of a hormone (i.e., whether it is water soluble or fat soluble) determines how it will travel through the bloodstream (alone or attached to a protein) and how it will bind to its receptor (fat-soluble hormones can travel through the membrane to receptors on the inside of the cell, while water-soluble hormones cannot pass through the membrane and must bind to receptors on the outside of the cell).

Steroid Hormones

Steroid hormones (including **estrogen**, **testosterone**, and **cortisol**) are fat-soluble molecules produced from cholesterol. Because they generally repel water, steroid hormones travel through the blood attached to carrier proteins. Once they reach their target cell, steroid hormones pass through the cell membrane and bind to receptors in the cytoplasm and genes in the nucleus to regulate protein production.

TABLE 4.1
Examples of Hormones within Each Class

Protein and peptide hormones	Antidiuretic hormone (ADH), follicle-stimulating hormone (FSH), glucagon, growth hormone (GH), insulin, luteinizing hormone (LH), oxytocin, prolactin (PRL), thyroid stimulating hormone (TSH), thyrotropin releasing hormone (TRH)
Steroid hormones	Aldosterone, cortisol, estrogen, testosterone
Amino acid derivatives	Epinephrine, norepinephrine, thyroxine, triiodothyronine

Amino Acid Derivatives

Amino acid derivatives (such as epinephrine and norepinephrine) are water-soluble molecules derived from amino acids (compounds that form proteins). These hormones travel freely in the blood, but they cannot pass through the cell membrane, so they bind to receptors on the surface. Binding activates second messengers inside the cell that trigger enzymes or influence gene expression.

Protein and Peptide Hormones

Protein and peptide hormones (including **insulin**, **prolactin**, and growth hormone) are water-soluble hormones made up of amino acid chains. Like amino acid derivatives, peptide hormones circulate alone and bind to receptors on the cell surface. Protein and peptide hormones consist of chains of amino acids. These chains may number only a few amino acids in length, as is the case with many peptide hormones (**thyrotropin-releasing hormone [TRH]** contains only three amino acids); or they may contain more than 200 amino acids, as do many protein hormones (**follicle-stimulating hormone [FSH]** contains 204 amino acids).

A **glycoprotein** is a special type of protein hormone consisting of a protein connected to a glucose (sugar) molecule. Examples include **luteinizing hormone (LH)**, follicle-stimulating hormone, and **thyroid-stimulating hormone (TSH)**.

Peptide and protein hormones are produced in the endocrine cell under the direction of mRNA (messenger ribonucleic acid). The mRNA contains information that dictates the amino acid sequence of the protein. mRNA originates in the cell nucleus, then moves out into the cytoplasm, where it serves as a template for the amino acids to form an inactive molecule called a **preprohormone** (or prohormone). The prohormone is packaged into a secretory granule, which carries it to the cell surface. When the granule meets the cell membrane, an enzyme processes the prohormone to release the active hormone from the cell into circulation.

Steroid hormones are synthesized from cholesterol, about 80 percent of which comes from food and is transported through the blood plasma as **high-density lipoprotein (HDL)** particles. Included among the steroid hormones are the sex steroids (estrogen, **androgens [testosterone]**, and **progesterone**) produced by the ovaries and testes, and the glucocorticoids (cortisol), **mineralocorticoids**, and androgens produced by the adrenal cortex.

To produce steroid hormones, enzymes convert cholesterol into a precursor molecule, called **pregnenolone**, in the cell mitochondria. Pregnenolone is then transported out of the mitochondria to the endoplasmic reticulum, where enzymes break it down further to produce either another precursor(s) or the active steroid hormone. Unlike protein hormones, which require granules to transfer them to the cell surface, steroid hormones can make the trip on their own and exit the cell via diffusion across the membrane.

Unlike protein and peptide hormones, which consist of several linked amino acids, amino acid derivatives contain just one or two amino acids. Two major groups of hormones, both derived from the amino acid **tyrosine**, fall within this category: thyroid hormones (**thyroxine [T4]** and **triiodothyronine [T3]**) and **catecholamines** (epinephrine and norepinephrine, which are both hormones and neurotransmitters). Tyrosine reaches the endocrine cell via the bloodstream. Once inside the cell, enzymes transform the tyrosine into the active hormone. In the case of thyroid hormones, iodine is added to the modified tyrosine molecules. Amino acid hormones, like steroid hormones, can travel on their own across the cell membrane.

Hormone Transport

Once a hormone is released from the cell, it travels through the bloodstream to the cell upon which it will act. To get there, a hormone may either circulate alone (free) or bound to a carrier protein in the blood. As mentioned above, amino acid, peptide, and protein hormones typically circulate free because they are water soluble; steroid and thyroid hormones, which are fat soluble, circulate bound to proteins. The advantage to binding is that the carrier protein helps the hormone navigate through all of the cellular traffic in the body to reach its target tissue. A bound hormone also stays longer in the blood than a free hormone, because its carrier protein holds it back from crossing a cell membrane. The level of bound and free proteins in the blood usually remains stable, because the proteins are in a concentration equilibrium, so each time a newly produced hormone is released into circulation, a hormone that is bound to a protein is freed.

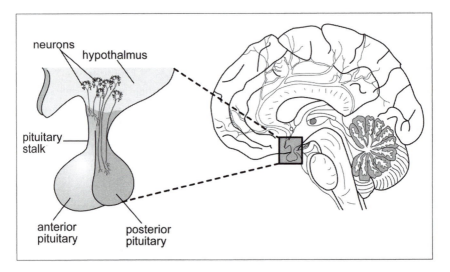

Figure 4.1   The Hypothalamus. Hormones released from the hypothalamus travel through blood vessels in the pituitary stalk to the anterior pituitary. (Sandy Windelspecht/Ricochet Productions)

Although most hormones travel through the bloodstream to reach their target cells, there are exceptions. One example is the **hypothalamichypophyseal portal system** (Figure 4.1), through which releasing hormones secreted by nerve cells in the **hypothalamus** travel (via capillaries in the hypothalamus and veins in the pituitary stalk) directly to the **anterior pituitary** without ever entering the general circulation.

Target Cells and Receptors

A hormone cannot trigger a physiological reaction in just any cell—it is specifically designed to act upon only those cells that are receptive to it, which are called **target cells**. How does a hormone find its target cell? Each target cell comes complete with **receptors**—proteins that lie either on the surface of the cell or within the nucleus. The receptors exhibit specificity and bind only to the right hormone—or hormones—from a sea of other molecules. A cell's response depends upon the concentration of the hormone and the number of receptors to that hormone that it contains.

When the hormone binds to its receptor, it initiates a chain of events that ultimately alters the cell's function. The activated hormone-receptor complex can have one of three main effects: It can instruct cells to either make or stop making RNA from DNA (by the process of **gene transcription**), thus starting or stopping protein production; it can turn enzymes in the cell on or off, thus altering the cell's metabolism; or it can change the permeability of the cell membrane to allow in or shut out certain chemicals. If an individual lacks receptors for a particular hormone (or hormones), that hormone will not be able to do its job, and disease will often result. Hormones not only trigger production of certain proteins within a cell; they may also block protein production or even block other hormones from binding to the cell receptor. Based on their effect, hormones are assigned to one of four classifications:

- *Agonists*: **Agonists** are hormones that bind to their receptor and elicit a specific biological response. For example, a glucocorticoid is an agonist for the receptor that binds cortisol.

- *Antagonists*: **Antagonists** are hormones that bind to the receptor but do not trigger a biological response. By occupying the receptor, the antagonist blocks an agonist from binding and thus prevents the triggering of the desired effect within the cell. For example, an antiandrogen is used to block the function of androgens in hormone therapy.

- *Partial agonist–partial antagonist*: A hormone that, when bound to the receptor, initiates a lesser biological response within the cell. By occupying the receptor, the partial agonist–partial antagonist blocks the potential action of an agonist, which could have generated a more significant biological response within the cell.

- *Mixed agonist–antagonist*: A hormone that exerts a different action on the receptor, acting as either an agonist or antagonist, depending upon the situation.

There are two types of receptors. Water-soluble hormones are unable to cross the membrane on their own because they are repelled by the fatty membrane that surrounds each cell, so they bind to receptors on the cell surface. Hormones that are fat soluble (such as steroids) are able to cross the membrane, so they bind to receptors inside the cell.

Cell Surface Receptors and Second-Messenger Systems

Glucagon, catecholamines, **parathyroid hormone (PTH)**, **adrenocorticotropic hormone (ACTH)**, thyroid-stimulating hormone (TSH), and luteinizing hormone (LH) are water soluble and therefore cannot cross into the cell. Instead, they bind to receptors on the surface of the cell membrane and trigger a cascade of events that leads to the desired biological response within the cell. For the message to pass from the hormone into the cell requires the efforts of second messengers (Figure 4.2), which activate enzymes or other molecules inside the cell.

The hormone's actions are similar to those of a witness to a car accident. The witness acts as a first messenger, calling 911 and alerting the

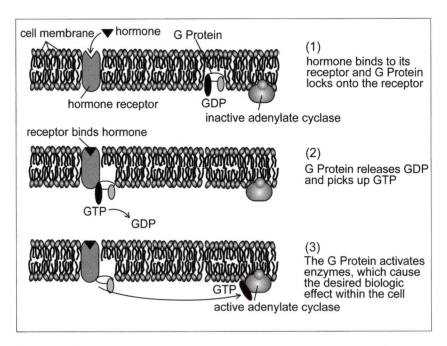

Figure 4.2   Second messenger system. (Sandy Windelspecht/Ricochet Productions)

operator to the problem. The 911 operator is the receptor, taking the message and passing it along to the emergency crew, which acts as the second messenger, coming to the aid of the accident victims.

One of the primary second-messenger systems involves G proteins. G proteins are like chemical "switches" that have to be triggered before the hormonal message can be passed into the cell. When a hormone binds to its receptor, the receptor changes shape and activates a G protein inside the cell. The G protein releases a guanine nucleotide it had been holding called guanine diphosphate (GDP), then grabs another, similar nucleotide called guanine triphosphate (GTP). Then the G protein goes to work, activating enzymes such as adenylyl cyclase. These enzymes produce a second messenger called cyclic adenosine monophosphate (cAMP). cAMP relays messages to effectors (molecules that regulate a series of chemical reactions) inside the cell, which lead to the desired biological reaction (for example, releasing glucose from cells when the body needs it for energy).

Intracellular Receptors

Receptors for steroid and thyroid hormones, as well as for vitamin D (a vitamin with hormone properties), are located inside the cell nucleus or cytoplasm. These hormones are fat soluble and can therefore cross the cell membrane on their own via simple diffusion. When they enter the cell, they meet up with and bind to their receptors, forming a hormone-receptor complex. The complex binds to parts of DNA in the cell nucleus called hormone response elements. Binding alters the DNA, resulting in the synthesis of a new protein.

Hormone Regulation and Secretion

Hormones are so potent that just a tiny amount can exert powerful influences throughout the body. If too much or too little of a hormone is in circulation, the body can fall prey to serious disease. The effects of a particular hormone are related to its concentration in the bloodstream. Concentration is affected by the rate of production, the speed of distribution to target cells, and the speed at which the hormone is degraded after it is released from its receptor. All of these elements are strictly controlled by feedback loops or mechanisms, which measure and respond to changes within the body. Feedback loops ensure that enough hormones are produced to complete necessary tasks and keep the endocrine system tightly integrated with the nervous and immune systems.

Some endocrine cells secrete their hormones at set times every day, every month, or even every year. Other cells secrete hormones following stimulation by other hormones, or in response to internal or external stimuli.

Some hormones are released in regular patterns that follow a 24-hour cycle (called circadian rhythms). Cortisol release, for example, rises in the early morning, gradually drops during the day, and stays very low during sleep. Other hormones follow a monthly, or even a seasonal, pattern. The pituitary gland, for example, releases luteinizing hormone and follicle-stimulating hormone in response to variations in a woman's monthly menstrual cycle.

Hormonal Influences

The majority of hormones are regulated by other hormones (called tropic hormones), which either stimulate or inhibit their release based on the body's needs. The hypothalamus-pituitary control system is an example of tropic influence. The hypothalamus secretes several neurohormones, which signal the pituitary to release its own hormones. Pituitary hormones, in turn, direct the functions of several target organs.

Internal and External Influences

Many endocrine glands have their own mechanisms for sensing whether they need to release hormones. The endocrine-producing islet cells of the pancreas detect glucose levels in the blood, and release or inhibit insulin production as necessary. Sometimes, external factors are involved in hormone secretion. When a baby nurses from its mother's breast, the suckling action stimulates secretion of the hormones prolactin and oxytocin. Prolactin causes milk production in the mammary glands and maintains lactation. Oxytocin stimulates the release, or let-down, of milk into the nipple.

Feedback Mechanisms

Hormones regulate one another through feedback loops. The most basic feedback systems involve only one closed loop. More complex systems consist of a series of interrelated loops. Two main types of feedback systems exist.

Negative Feedback

The most common type of feedback works much like a home air conditioning unit. When the temperature in the home rises to a preset level, a sensing mechanism turns on the air conditioner. After the air conditioning has run long enough to drop the temperature to a comfortable level, the shut-off mechanism is activated. Thanks to the feedback mechanism, the home is never allowed to get too cold or too warm.

In the body's negative feedback loop, a physiological change triggers the release of a particular hormone. Once the level of this hormone rises in the blood, it signals the endocrine cells that secreted it to stop producing it. Negative feedback prevents the overproduction of hormones, which could lead to disease.

An example of a simple negative feedback loop occurs after a person eats a piece of cake. After the cake is ingested, glucose (sugar) levels in the blood rise. In response to rising glucose levels, the endocrine cells of the pancreas release insulin. Insulin helps the cells take in and use glucose, lowering the amount of the sugar in the bloodstream. When blood glucose levels fall back to a normal level, insulin release is inhibited.

Positive Feedback

As the name implies, a positive feedback loop stimulates, rather than inhibits, the production of a particular hormone. One example involves the release of oxytocin from the pituitary gland during childbirth. Oxytocin stimulates uterine contractions, which help push the baby out of the uterus. As levels of oxytocin in the blood rise, they trigger the pituitary to secrete even more of the hormone. Uterine contractions continue to increase until the child is finally born. Positive feedback is far less common than negative feedback, because it has the potential to contribute to dangerously high hormone levels.

More complex feedback systems involve several interrelated loops. One example is the hypothalamic-pituitary-target organ axis, a multi-loop system that coordinates the efforts of the hypothalamus in the brain, the pituitary gland, and the target gland.

The hypothalamus, in response to reduced hormone levels in the bloodstream, stimulates pituitary hormone secretion. The pituitary hormone travels through the bloodstream to act on its target tissue(s). As the level of pituitary hormone in the blood rises, the hypothalamus stops secreting its releasing hormone. Consequently, the pituitary stops producing its own hormone, and blood levels of the hormone return to normal.

The hypothalamus has the ability to override the system, increasing or reducing hormone levels to adjust to physical and emotional stresses. Hormones from the target gland bind to nerve cells in the hypothalamus, which inhibit or trigger production of releasing hormones that influence pituitary hormone secretion. Without this mechanism, hormone levels would remain constant, even when they were needed in greater amounts to mediate a stress response.

Hormone Elimination

After hormones interact with their target cells and produce the desired result, they are no longer needed by the body. Most hormones are either converted to less active molecules or degraded by enzymes into an inactive form before being excreted in the urine or feces. Very few hormones are eliminated intact. Peptide hormones, catecholamines, and eicosanoids are all degraded by enzymes in the cell. Steroid hormones are metabolized into inactive forms and eliminated by the kidneys.

Endocrine Glands: Anatomy and Function

The endocrine system is made up of a complex network of glands (Figure 4.3), each of which secretes hormones that coordinate and regulate functions throughout the body. The pituitary, thyroid, parathyroids, adrenals, gonads (testes and ovaries), and endocrine pancreas are considered traditional endocrine glands with the primary purpose of secreting hormones into the bloodstream. But other, nonendocrine organs—including the heart, brain, kidneys, liver, skin, and gastrointestinal tract—can also secrete hormones.

By definition, endocrine glands release their hormones into the bloodstream to act upon target tissues elsewhere in the body. Endocrine glands should not be confused with **exocrine** glands, which secrete their substances to the outside of the body, to internal cavities such as the lumen of the

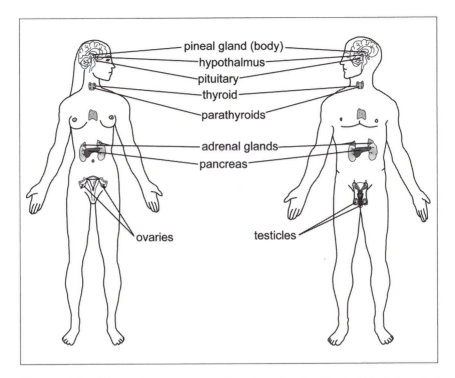

Figure 4.3 The endocrine system: female and male. (Sandy Windelspecht/ Ricochet Productions)

intestines, or to other tissues through ducts (for example, the salivary and sweat glands).

The endocrine system is assigned several critical responsibilities, the most important of which are to maintain a constant internal environment (**homeostasis**); aid in growth, development, reproduction, and metabolism; and coordinate with the central nervous and immune systems.

Water and Electrolyte Balance

For the body to function properly, it needs to maintain an internal balance of fluids and **electrolytes** (electrically charged chemical ions such as sodium, potassium, chloride, calcium, magnesium, and phosphate). More than 40 quarts (37 liters) of water circulate throughout the body. About two-thirds is **intracellular** fluid, located within the cells. About 75 percent of the remaining extracellular fluid is found in the tissue outside of the cells, and the other 25 percent is contained within the fluid portion of blood (plasma).

A rise in blood fluid volume (overhydration) can force the heart to work harder and dilute essential chemicals in the system. Too little water, or dehydration, can lead to low blood pressure, shock, and even death. The kidneys help to balance the fluid in the body by reabsorbing liquid into the bloodstream when levels get too low, or by eliminating excess fluids when levels rise too high. These processes occur under the direction of the endocrine system.

If the concentration of water drops too low (because not enough liquid was ingested or because fluid was lost through sweating, vomiting, or diarrhea), neurons called **osmoreceptors** send a message to the hypothalamus in the brain, which in turn tells the pituitary gland to secrete **antidiuretic hormone (ADH)** (also known as **vasopressin**) into the bloodstream. This hormone increases the permeability of the distal convoluted tubules and the collecting ducts in the nephrons of the kidneys, thus returning more fluid to the bloodstream. When more water is reabsorbed, the urine becomes more highly concentrated and is excreted in smaller volume. When the fluid concentration in the body is too high, ADH is not released. The distal convoluted tubules and collecting ducts are less permeable to water, and the kidneys filter out excess fluid, producing a larger volume of more dilute urine.

The kidneys must also maintain a balance of sodium, potassium, and other electrolytes in body fluids. To do this, they separate ions from the blood during filtration, returning what is needed to the bloodstream and sending any excess to the urine for excretion. Electrolyte levels are also directed by the endocrine system.

Sodium and potassium are two of the most important electrolytes, because without them, fluids would not be able to properly move between the intracellular and extracellular spaces. Sodium is the most abundant electrolyte in the extracellular fluid, and it also plays an important role in nerve and muscle function. The presence of too much sodium (a condition called hypernatremia) will send water from inside the cells into the extracellular region to restore balance, causing the cells to shrink. If nerve cells are affected, the result can be seizures and, in rare cases, coma. Too little sodium (called hyponatremia)—lost from excessive diarrhea, vomiting, or sweating—can send water into the cells, causing them to swell. Hyponatremia can lead to weakness, abdominal cramps, nausea, vomiting, or diarrhea. The swelling is even more dangerous if it occurs in the brain, where it can cause disorientation, convulsions, or coma.

Potassium assists in protein synthesis and is crucial for nerve and muscle function. Too little potassium can lead to a buildup of toxic substances in the cells that would normally pass into the extracellular fluid. To prevent a sodium-potassium imbalance, the cells use a mechanism called the sodium-potassium pump. This pump is a form of active transport (as opposed to the passive transport used in osmosis), which means that fluid can pass from one side of a semipermeable membrane to another, even if the concentration is already high on that side. But active transport requires energy to push molecules across the membrane. That energy is derived from adenosine triphosphate (ATP), a by-product of cellular respiration.

Once activated by ATP, the sodium-potassium pump pushes potassium ions into the cell while pumping sodium ions out of the cell until a balance is reached. Endocrine hormones regulate the amount of sodium and potassium in the bloodstream. In the case of a sodium imbalance, an enzyme secreted by the kidneys, called renin, stimulates the production of the hormone aldosterone by the adrenal glands, which are located just above the kidneys. Aldosterone forces the distal convoluted tubules and collecting ducts in the kidneys to reabsorb more sodium into the blood. It also maintains potassium homeostasis by stimulating the secretion of potassium by the distal convoluted tubule and collecting ducts when levels in the bloodstream get too high.

Parathyroid hormone (PTH), produced by the parathyroid glands, regulates levels of bone-building calcium and phosphate. When calcium concentrations in the body drop, PTH pulls calcium from the bones, triggers the renal tubules to release more calcium into the bloodstream, and increases the absorption of dietary calcium from the small intestine. When too much calcium circulates in the blood, the thyroid gland produces another hormone, **calcitonin**, which causes bone cells to pull more calcium from the blood, and increases calcium excretion by the kidneys. PTH decreases phosphate levels in the blood by inhibiting reabsorption in the kidney tubules, and calcitonin stimulates the bones to absorb more phosphate.

The following sections will provide a closer look at the function of each of the endocrine glands.

The Hypothalamus and Pituitary Gland

The hypothalamus and pituitary gland together serve as the command center of the endocrine system, and the core of the relationship between the endocrine and nervous systems. Together, they regulate virtually every physiological activity in the body. As mentioned earlier, the nervous and endocrine systems also regulate each other: neurohormones from the hypothalamus direct the release of endocrine hormones, and hormones from the endocrine system regulate nervous system activity.

The Hypothalamus

The tiny, cone-shaped region at the base of the brain (Figure 4.4) called the hypothalamus coordinates the neuroendocrine system, helps regulate metabolism, and controls the part of the nervous system that oversees a number of involuntary bodily functions (sleep, appetite, body temperature, hunger, and thirst). It also serves as the link between the nervous and endocrine systems.

The hypothalamus projects downward, ending at the pituitary stalk, which connects it to the pituitary gland. Together, the hypothalamus and pituitary (known collectively as the hypothalamic-pituitary axis [HPA]) direct the functions of the endocrine system. Although the pituitary has been termed the "master gland," the hypothalamus is the real control center behind the operation. The hypothalamus sends out messages (releasing or inhibiting hormones), which signal the pituitary to release—or stop releasing—its hormones. Pituitary hormones control the functions of virtually every endocrine gland in the body.

The hypothalamus is made up of clusters of **neurosecretory cells**, which both transmit electrical messages (impulses) and secrete hormones. Electrical impulses are transmitted from one nerve cell to another via chemical messengers called neurotransmitters. The impulses travel across junctions called synapses and bind to receptors on the receiving neuron. Neurotransmitters are chemical compounds that are made up of simple or more complex amino acid sequences or peptides. Examples of

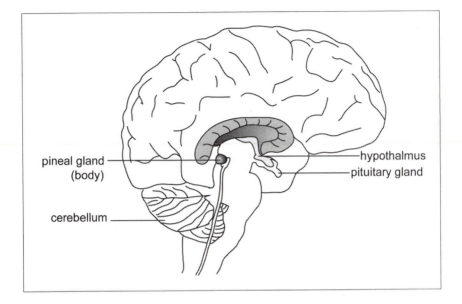

Figure 4.4 The Pineal Glands. The hypothalamus, pituitary, and pineal body. (Sandy Windelspecht/Ricochet Productions)

neurotransmitters include epinephrine, norepinephrine, serotonin, acetylcholine, dopamine, and histamine.

The hypothalamus also secretes a number of hormones that are referred to as **neurohormones**, which either travel through the body via the general circulation or go directly to the anterior pituitary gland through a portal network of blood vessels (called the hypothalamic-hypophyseal portal system) and signal it to release or stop releasing its hormones. Hypothalamic hormones are called releasing or inhibiting hormones, depending upon how they influence the pituitary gland. As their names suggest, releasing hormones trigger hormone secretion, and inhibiting hormones halt hormone secretion.

One of the neurohormones is the growth hormone-releasing hormone (GHRH), which is a large peptide hormone that stimulates the secretion of growth hormone from the anterior pituitary gland. Release of GHRH is triggered by stress (such as exercise) and is inhibited by somatostatin, which is also released by the hypothalamus. Negative feedback is largely controlled by compounds known as somatomedins, growth-promoting hormones made when tissues are exposed to growth hormone.

The Pituitary Gland

The pea-shaped pituitary gland (also known as the **hypophsis**) sits nestled in a cradle of bone at the base of the skull called the sella turcica ("Turkish saddle"). It is attached to the hypothalamus by the pituitary (hypophyseal) stalk, through which run the blood vessels and nerves (axons) that deliver hypothalamic hormones to the anterior pituitary.

As previously noted, the pituitary (Figures 4.5 and 4.6) is often referred to as the "master gland" because it directs the functions of most other endocrine glands (including the adrenals, thyroid, and gonads [ovaries and testes]). In addition to stimulating other endocrine glands to release their hormones, the pituitary secretes several of its own hormones: growth hormone (GH), prolactin, and oxytocin.

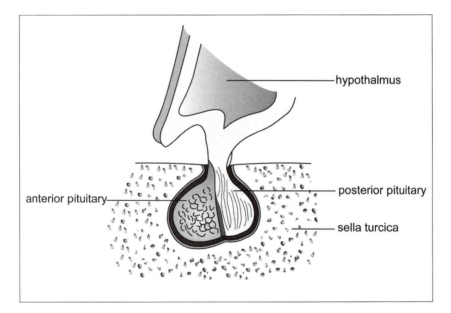

Figure 4.5   The anterior and posterior pituitary gland. (Sandy Windelspecht/ Ricochet Productions)

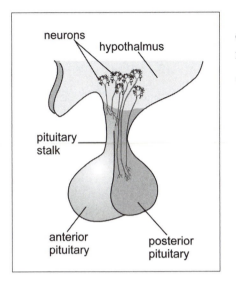

Figure 4.6 The pituitary gland. (Sandy Windelspecht/Ricochet Productions)

The pituitary gland is made up of three lobes: the anterior, intermediate, and posterior.

- *Anterior pituitary*: The anterior pituitary is composed of endocrine cells, which secrete hormones in response to stimulation by hypothalamic hormones. Anterior pituitary hormones, in turn, stimulate the adrenal glands (**adrenocorticotropic hormone [ACTH]**), thyroid gland (thyroid-stimulating hormone [TSH]), and ovaries and testes (follicle-stimulating hormone [FSH] and luteinizing hormone [LH]). The anterior pituitary also produces growth hormone (which stimulates growth of bone and muscle) and prolactin (which initiates milk production following childbirth).

- *Intermediate pituitary*: The **intermediate pituitary** exists as a separate entity in animals, but only vestiges of this lobe remain in humans. Cells within the intermediate pituitary produce melanocyte-stimulating hormone, which controls skin pigmentation.

- *Posterior pituitary*: Although the **posterior pituitary** is situated next to the anterior pituitary, it has very different functions. The posterior pituitary is an extension of the nervous system, made up primarily of axons and nerve endings that reach down from the hypothalamus. The posterior pituitary stores and releases hormones that are actually produced within the hypothalamus: antidiuretic hormone (ADH), which helps the body conserve water by increasing reabsorption in the kidney tubules; and oxytocin, which stimulates uterine contractions during childbirth and triggers the letdown of milk from the mother's breast when her infant nurses.

The pituitary gland is divided into two separate units: the anterior pituitary and posterior pituitary, each of which functions independently and secretes its own set of hormones.

The Anterior Pituitary

The anterior pituitary is made up of five different types of cells, each of which secretes one or more different hormones:

- *Thyrotroph*: Thyroid-stimulating hormone (TSH)

- *Gonadotroph*: Luteinizing hormone (LH) and follicle-stimulating hormone (FSH)

- *Corticotroph*: Corticotropin (ACTH)

- *Somatotroph*: Growth hormone

- *Lactotroph*: Prolactin

Hormones are synthesized in the cytoplasm of the cell as larger, inactive molecules called prohormones. Neurohormones from the hypothalamus travel to the anterior pituitary via a closed system of veins (the hypothalamic-hypophyseal portal system) and signal the anterior pituitary to either release or stop releasing its hormones. When the signal is to release hormones, the hormone is activated from the prohormone as it is sent out from the cell into circulation. If the hypothalamus were destroyed, the anterior pituitary would be unable to secrete any of its hormones, with the exception of prolactin, which the hypothalamus primarily inhibits.

The anterior pituitary affects growth and metabolism in most other endocrine glands, as well as in other areas of the body. It also stimulates other endocrine glands to produce and secrete their hormones:

Thyroid-stimulating Hormone (TSH)

TSH, also called thyrotropin, is a large glycoprotein that affects cell growth and metabolism in the thyroid gland, and signals the gland to produce and release its hormones thyroxine (T4) and triiodothyronine (T3). Thyrotroph cells in the anterior pituitary release TSH after being stimulated by TRH from the hypothalamus. TSH release is inhibited by negative feedback involving thyroid hormones. When blood levels of thyroid hormones are high, somatostatin inhibits the production of TRH from the hypothalamus, which then inhibits TSH release. Glucocorticoids and estrogens also serve an inhibitory function by making the pituitary less responsive to TRH.

Gonadotropins

Luteinizing hormone (LH) and follicle-stimulating hormone (FSH) are **gonadotropins** secreted by cells called gonadotrophs in the anterior pituitary. These hormones promote egg and sperm development and control gonadocorticoid hormone (androgens and estrogen) release by the ovaries and testes. The secretion of both hormones remains very low from shortly after birth until puberty, when levels rise dramatically. In females, the rate of secretion varies at different times during the menstrual cycle. Secretion of both LH and FSH is controlled by gonadotropin-releasing hormone (GnRH) from the hypothalamus.

Luteinizing hormone: This hormone was given its name because it stimulates the conversion of the ovarian follicle to the **corpus luteum** following egg release. In the middle of a woman's menstrual cycle, estrogen levels rise, causing the release of GnRH from the hypothalamus. GnRH causes LH levels to surge, triggering ovulation. That surge allows the egg to rupture from its follicle and travel down the fallopian tube toward the uterus. Once the egg has been released, LH stimulates the conversion of the ovarian follicle to the corpus luteum, which produces progesterone, a hormone necessary to maintain pregnancy. LH also stimulates estrogen and progesterone production. In men, the hormone stimulates the growth of—and testosterone production in—the Leydig cells of the testes.

Follicle-stimulating hormone: In women, FSH stimulates the maturation of the ovarian follicles, in which the eggs develop. With the help of LH, FSH also increases estrogen secretion by the ovaries. In men, FSH acts upon the **Sertoli cells** (cells that line the seminiferous tubules, which nourish the germ cells from which sperm develop) of the testes, facilitating sperm maturation and development. Like LH, the release of FSH is stimulated by GnRH from the hypothalamus. FSH is inhibited by the hormone **inhibin**, which is secreted by the ovaries and testes.

Adrenocorticotropic Hormone (ACTH)

ACTH, also called corticotropin, is a small peptide hormone that stimulates cell development and hormone synthesis in the adrenal cortex (glucocorticoids, mineralocorticoids, and gonadocorticoids). In the fetus, ACTH also stimulates secretion of an estrogen precursor called dehydroepiandrosterone sulfate (DHEA-S), which prepares the mother for labor. Stress stimulates the release of CRH from the hypothalamus. CRH then activates the secretion of ACTH. ADH (vasopressin) also plays a role in ACTH release. High circulating levels of cortisol in the blood inhibit ACTH in two ways: by directly suppressing ACTH synthesis and secretion in the pituitary gland, and by acting on the hypothalamus to decrease CRH release. ACTH may also inhibit its own secretion.

Growth Hormone

Growth hormone, also called somatotropin, is a large polypeptide hormone produced by somatotroph cells in the anterior pituitary that plays a significant role in growth and metabolism. It primarily affects bone, muscle, and tissue growth. Without sufficient growth hormone, an individual would suffer from short stature. Too much growth hormone would result in gigantism. For normal growth to occur, the body requires energy, which growth hormone provides through protein synthesis and the breakdown of fats. Growth hormone has two types of effects: direct and indirect.

> *Direct effects*: Growth hormone acts directly upon protein metabolism, fat metabolism, and carbohydrate metabolism to help the body more efficiently use and conserve energy. It moves amino acids from the blood into cells and stimulates protein synthesis within the cells; it moves fats out of storage (in adipose tissue) for use in energy production to conserve proteins; and it decreases carbohydrate use and impairs glucose uptake into cells, thus sparing glucose for the brain.

Indirect effects: Growth hormone stimulates bone, muscle, and cartilage growth indirectly, by triggering the production of **insulin-like growth factor** 1 (IGF-1, or somatomedin). Insulin-like growth factors are synthesized in the liver and other tissues and act much like insulin, stimulating glucose uptake by cells. They also influence protein and DNA synthesis. IGF-1 stimulates proliferation of cartilage cells (called **chondrocytes**), causes muscle cell differentiation and proliferation, and initiates protein synthesis in muscle tissues.

Regulation of Anterior Pituitary Hormones

Hormones from the anterior pituitary can be regulated in one of three ways:

1. Hormones such as LH and FSH are released in pulses that follow a regular cycle. The strength and frequency of these pulses is set, in part, by the hypothalamus. Pulsatile release may follow a daily rhythm (circadian), or it may occur more frequently (ultradian) or less frequently (infradian) than once a day.

2. Most hormones are regulated by feedback loops, in which circulating hormone levels act upon the hormones that triggered their release. Three types of feedback loops exist:

   • *Long-loop system*: After being stimulated by a releasing hormone from the hypothalamus (CRH), the anterior pituitary signals its target organ (for example, the adrenal cortex) to produce its hormone (cortisol). When that hormone reaches a certain level in the system, it acts upon the hypothalamus via negative feedback, inhibiting its releasing hormone (CRH). When the hypothalamus stops or decreases production of the releasing hormone, the anterior pituitary subsequently stops releasing its hormone (ACTH).

   • *Short-loop system*: Some hormones (for example, LH and FSH) can suppress their own release without entering the bloodstream.

   • *Ultrashort-loop system*: A releasing hormone (for example, LHRH or GHRH) can act directly on the hypothalamus to regulate its own secretion.

3. Finally, hormone release can be influenced by external factors, such as stress (for example, the fight-or-flight release of corticotropin from the adrenal cortex), diet, and illness.

The Posterior Pituitary

The posterior pituitary is not a classic endocrine organ because it is composed primarily of extensions of axons and nerve endings from the hypothalamus. Its two hormones, antidiuretic hormone (ADH, or vasopressin) and oxytocin, are actually produced in the neurons of nuclei in the hypothalamus. These hormones travel down nerve fibers to the posterior pituitary, which merely stores and releases them.

Antidiuretic Hormone

Antidiuretic hormone (ADH, or vasopressin) is a small peptide hormone whose primary role is to conserve water in the body by signaling the kidneys to excrete less fluid. The hormone is synthesized as a preprohormone in the hypothalamic neurons. The preprohormone is converted into a prohormone, which contains an attached protein called neurophysin that is removed as the hormone is secreted.

The human body contains about 60 percent water. Significant fluctuations in water balance (i.e., dehydration or overhydration) can be extremely dangerous to the system. When the concentration of liquid in the blood drops, special sensors (called osmoreceptors) in the hypothalamus alert the neurons that produce ADH. Osmoreceptors are extremely sensitive and can respond to tiny changes (as small as 1 percent) in water concentration.

Whereas a diuretic increases urine output, an antidiuretic conserves fluid in the body by reducing urine output. When ADH is released, it binds to receptors in the distal or collecting tubules of the kidneys and increases their permeability, thus stimulating the reabsorption of liquid back into the blood (normally, these tubules are virtually impermeable to water). When more water is absorbed into the bloodstream, blood volume and pressure increase. Conversely, when the fluid level in the body gets too high, ADH release is suppressed, and the kidneys excrete more liquid into the urine. ADH also constricts blood vessels (vasoconstriction), the role for which it was given its alternate name, vasopressin.

Thirst, the body's physical indicator that fluid levels are low, is also regulated by osmoreceptors in the hypothalamus, although not the same ones that trigger ADH release. The body first sends in ADH to try to regulate water balance; then, if that measure fails to increase fluid volume, it invokes thirst. Changes in blood pressure also stimulate ADH release. Two pressure sensors—one in the carotid artery in the neck, and the other in specialized cells in the atrium of the heart—discern changes in blood pressure and volume. They send a message, via nerves, to the hypothalamus. When these sensors are stretched by expanding blood volume, they shut off ADH secretion so that more water is excreted by the kidneys. When they sense reduced blood volume (for example, when a person is hemorrhaging from a severe injury), they trigger ADH production.

Oxytocin

Oxytocin is similar in structure to ADH, and it is also synthesized from preprohormones in the hypothalamus and transported to the posterior pituitary for secretion into the blood. In addition to being secreted by the pituitary, oxytocin is also released by tissues in the ovaries and testes. The primary function of oxytocin is to stimulate the mammary glands in a mother's breast during lactation—to let down the milk so that her baby can nurse. Oxytocin (which is derived from the Greek word meaning "swift birth") also stimulates uterine contractions during labor. The release of oxytocin into a new mother's brain also helps forge a bond between her and her new baby. The hormone normally circulates in low levels in both men and women, but it rises in women during ovulation, birth, and lactation, as well as in times of stress.

Pineal Gland

The small, cone-shaped pineal gland was once called the "third eye" and ascribed supernatural powers. It extends downward from the third ventricle of the brain, above and behind the pituitary gland. The pineal gland is composed of parts of neurons, but otherwise has no direct neural connection with the brain.

Scientists know very little about the gland and what it does, but they do know that it secretes the hormone melatonin, which responds to light and dark, and communicates that information to the rest of the body. Melatonin influences circadian rhythms (the body's daily biological clock) and thus plays a role in functions regulated by night/day cycles, including reproduction and sleep/wake patterns.

Melatonin synthesis begins with the amino acid tryptophan, which is then converted into serotonin and finally into melatonin. Melatonin release is regulated by the sympathetic nervous system and is stimulated primarily by darkness, but it can also be triggered by hypoglycemia (low blood sugar). Melatonin concentration is highest at night and falls to almost undetectable levels during the day.

Although its function is still not completely clear, melatonin is believed to act upon suprachiasmatic nuclei (tightly packed groups of small cells) in the hypothalamus (which have receptors to it) to influence the body's daily biological rhythms. Synthetic versions of the hormone have been used to treat everything from jet lag to insomnia.

The Thyroid and Parathyroid Glands

The thyroid and parathyroid glands in the neck have several life-sustaining functions: The thyroid gland produces hormones that affect growth, development, metabolism, calcium homeostasis, and cell differentiation; and the parathyroid glands regulate calcium and phosphorous levels.

The Thyroid Gland

The largest endocrine gland in the body sits just below the larynx (voice box) and wraps around the trachea (windpipe). The thyroid gland (Figure 4.7) resembles a butterfly, with its two lobes reaching out like wings on either side of a narrow strip of tissue called the isthmus. In a healthy adult, the thyroid weighs about 20 grams, but it can grow to several times this size.

The thyroid gland is the largest endocrine organ and is crucial to nearly all of the body's physiological processes. It produces thyroid hormones, which are needed for growth, development, and a variety of metabolic activities. The thyroid is composed of two types of cells: follicular and parafollicular.

Follicles are sacs filled with the prohormone thyroglobulin. Thyroglobulin breaks apart to produce the two thyroid hormones, thyroxine (T4)

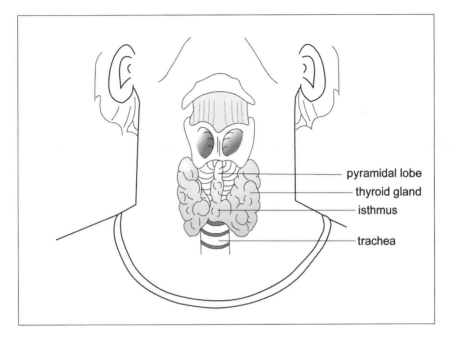

Figure 4.7   The thyroid gland. (Sandy Windelspecht/Ricochet Productions)

and triiodothyronine (T3). Lining the sacs are follicular cells, which synthesize and then either secrete or store these hormones. Parafollicular (or C) cells fill the spaces in between follicles. They secrete the hormone calcitonin. Inside the follicles is a substance called colloid, which consists primarily of the glycoprotein thyroglobulin.

Thyroid Hormone Synthesis

What makes the thyroid gland unusual among endocrine organs is that it requires iodine to produce its hormones. Iodine enters the body through food (i.e., iodized salt and bread) and water in the form of iodide or iodate ion.

The thyroid gland captures this iodide, and the enzyme thyroid peroxidase activates it. The follicular cells produce thyroglobulin, which is deposited in the colloid. Tyrosine is bound to the thyroglobulin molecule. Iodine diffuses into the colloid and is added to the thyroglobulin. Enzymes break down the thyroglobulin, releasing thyroid hormones into the bloodstream.

Without sufficient iodine, the thyroid reduces its hormone output. Decreased levels of thyroid hormones in the blood stimulate the anterior pituitary to secrete more thyroid-stimulating hormone (TSH) to make up for the deficit. The thyroid gland swells in size as it tries to increase its output, a condition called goiter.

Unlike most other endocrine organs, which produce and immediately secrete their hormones, the thyroid can store its hormones for several weeks. It releases its hormones when acted upon by TSH from the pituitary gland. TSH stimulation leads to the activation of thyroxine (T4). It splits from the thyroglobulin molecule as it leaves the cell and enters the bloodstream. T4 is the more plentiful of the two hormones, making up about 90 percent of the total thyroid hormones. Triiodothyronine (T3) is secreted in much smaller concentrations than thyroxine, but it is the more active of the two hormones. T3 is produced (usually in the peripheral tissues, especially the liver and kidney) when thyroxine loses one of its iodine molecules.

Because thyroid hormones are not water soluble, they generally travel through the bloodstream attached to carrier proteins (most often thyroxine-binding globulin, but also to a lesser extent to thyroxine-binding prealbumin and albumin). Virtually every cell in the body contains thyroid hormone receptors. Once the thyroid hormone reaches its target cells, it travels through the membrane via diffusion or with the help of carriers, and it binds to receptors in the nucleus. Thyroid hormones affect gene transcription, which either stimulates or inhibits protein synthesis.

Regulation of Thyroid Hormone Secretion

The hypothalamic-pituitary axis is the primary regulator of thyroid hormone production and secretion (Figure 4.8). Thyrotropin-releasing hormone (TRH) from the hypothalamus travels to the anterior pituitary and stimulates it to release thyroid-stimulating hormone (TSH). TSH influences every step of the thyroid production process, from thyroid cell growth to iodide uptake and metabolism. Finally, TSH triggers hormone secretion.

TSH release is stimulated and inhibited by positive and negative feedback, based on circulating levels of T4 and T3. When these hormones are in short supply, the pituitary and hypothalamus act on the thyroid to increase its production. Conversely, when too much of these hormones circulate in

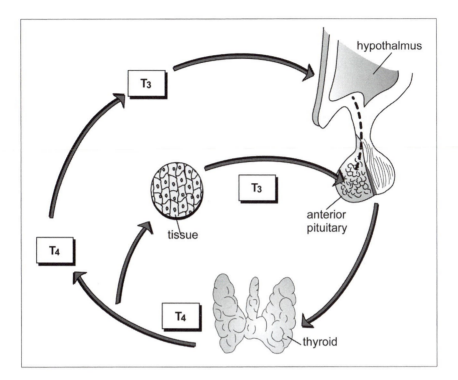

Figure 4.8   Thyroid Hormone Regulation. (1) TRH from the hypothalamus triggers TSH secretion from the pituitary, (2) TSH stimulates thyroid hormone release, and (3) circulating levels of T4 and T3 influence the hypothalamus and pituitary to regulate thyroid production. (Sandy Windelspecht/Ricochet Productions)

the blood, the pituitary and hypothalamus slow or stop thyroid production. Feedback ensures that hormone levels remain at an appropriate level.

Also controlling thyroid hormone secretion is the conversion of T4 into T3. As more T4 is converted into T3, rising T3 levels make the pituitary less responsive to TRH. But as more T4 is lost because of this conversion, the pituitary once again becomes more sensitive to TRH stimulation.

The Parathyroid Glands

Most healthy adults have two pairs of oval-shaped parathyroid glands, which lie next to the thyroid gland in the neck (the word "parathyroid" means "beside the thyroid"). In some instances, individuals may have fewer than or more than four parathyroid glands. Inside the glands are clusters of epithelial cells that produce and secrete parathyroid hormone, which is the most significant regulator of calcium levels in the blood. Calcium is essential for cell function as well as for bone formation.

The parathyroid glands produce only one major hormone: parathyroid hormone (PTH). PTH (also called parathormone), a polypeptide, opposes the actions of calcitonin by increasing blood calcium levels. Without this hormone, calcium concentrations in the blood would drop to life-threatening levels (**hypocalcemia**). Another effect of PTH is to reduce blood levels of phosphorous.

The parathyroids synthesize PTH from a larger, inactive prohormone in the parenchymal parathyroid cell. As PTH is released from the cell, it is split from the prohormone. PTH increases blood calcium by acting upon the bones, the kidneys, and (indirectly) the small intestine:

- *Bone*: PTH releases calcium from bone by stimulating the formation and activity of bone-dissolving cells called osteoclasts.

- *Kidneys*: PTH increases calcium reabsorption in the kidney tubules, reducing the amount of calcium that is lost in the urine. Because the kidneys filter a large volume of calcium each day, even a slight adjustment in excreted calcium can have a big effect on body chemistry. PTH also decreases phosphorous reabsorption, so the kidneys excrete a greater amount.

- *Small Intestine*: The effects of PTH on the small intestine are indirect. PTH increases production of vitamin D metabolites (the active form of vitamin D) in the kidneys (more on vitamin D later in this chapter). These metabolites increase the rate at which ingested calcium is absorbed in the small intestine, providing more calcium to circulate in the bloodstream.

PTH Regulation

Because blood calcium levels are so crucial to normal body function, cells of the parathyroid contain special receptors that can sense minute changes in calcium concentration. When calcium binds to these receptors, it results in reduced PTH secretion, which lowers calcium concentration in the blood. Without the influence of bound calcium, the receptors continue to stimulate PTH secretion. In the event that blood calcium levels remain depressed, PTH secretion can increase to 50 times its normal levels.

Parathyroid Hormone-related Protein (PTHrP)

PTHrP is similar in structure and function to parathyroid hormone (it too affects calcium and phosphorous balance), but it is produced in many tissues throughout the body. PTHrP binds to PTH receptors and has several of its own receptors as well. PTHrP can either act upon cells in other parts of the body or influence the nucleus of the cell(s) in which it was produced (called intracrine action). Like PTH, it releases calcium from bone into the blood and increases reabsorption in the kidneys.

The Adrenal Glands and Endocrine Pancreas

The adrenal glands are often referred to as the fight-or-flight glands because they secrete hormones involved in the body's stress response (Figure 4.9). The small, triangular glands sit on top of each kidney, surrounded by a capsule of connective tissue. When the body is confronted with stress—a serious car accident or a career-ending confrontation with one's boss, for example—the two adrenal glands kick into gear. The substances they produce (the catecholamines from the adrenal medulla and the steroid hormones from the adrenal cortex) orchestrate the stress response, making the body more alert, more energy efficient, and ready to face the daunting task at hand. Adrenal hormones are also involved in a number of other functions: regulating electrolyte balance, blood sugar levels, and metabolism; and influencing sexual characteristics.

Like the pituitary gland, the adrenals are essentially two glands in one: an outer cortex and an inner medulla. Each region originates from a separate embryological source, functions separately, and produces its own distinct hormones.

Adrenal Medulla

Hormones produced in the central, medullary region of the adrenal gland are referred to as the catecholamines. They are both hormones and

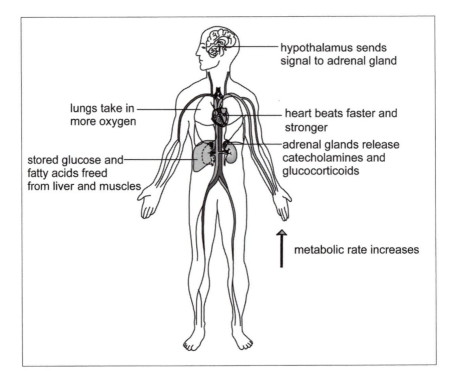

Figure 4.9 The fight-or-flight response. (Sandy Windelspecht/Ricochet Productions)

neurotransmitters, because they are produced and secreted by sympathetic nerves (including neurons in the brain). The primary catecholamines are epinephrine (adrenaline), norepinephrine (noradrenaline), and dopamine.

Catecholamines are produced in a multistep process that begins with the amino acid tyrosine. First, an enzyme converts tyrosine into the chemical L-dopa. A second enzyme converts L-dopa to dopamine. Dopamine is then converted to norepinephrine by yet another enzyme. Finally, epinephrine is synthesized from norepinephrine before being released into the blood. Epinephrine makes up the bulk (80 percent) of catecholamines released from the adrenal medulla; the remaining 20 percent is norepinephrine.

Catecholamine release is part of what Harvard physiologist Walter Cannon (1871–1945) termed the body's fight-or-flight response. When confronted with stress, the body shifts into overdrive. It becomes more alert and refocuses energy as it prepares to either stay and fight the danger or run away as quickly as possible.

As soon as the physical or emotional trauma occurs, the hypothalamus sounds the alarm. It sends out nerve impulses, which race to the adrenal medulla and signal it to release epinephrine and norepinephrine. These substances course through the blood attached to carrier proteins such as albumin, and they bind to adrenergic receptors on the surface of their target cells. Two types of adrenergic receptors exist: alpha-adrenergic and beta-adrenergic. The response elicited depends on the type of receptor stimulated. Alpha receptors are involved in smooth muscle contractions, pupil dilation, and blood vessel contraction. Beta receptors stimulate the heart and lungs, and relax the uterus.

Once the hormones are bound to their receptors, the body undergoes a rapid and dramatic transformation. The heart beats faster and stronger, the pupils dilate, the skin breaks out in a sweat, and the breathing becomes more intense as the body reaches a new level of alertness. Under the surface, a number of important physiological changes occur, directed by the adrenal catecholamines:

- The heart beats faster and more forcefully (primarily as a result of epinephrine), rushing additional blood throughout the body (especially to the brain and muscles). At the same time, the arteries constrict (primarily as a result of norepinephrine), increasing blood pressure.

- Stored glucose and fatty acids are freed to be used for energy. Catecholamines release glucose from the liver and muscles by stimulating the breakdown (**glycogenolysis**) of glycogen—the stored form of glucose. They also release fatty acids from adipose tissue by breaking down fatty compounds called triglycerides. The catecholamines oppose the action of insulin by preventing glucose movement into muscle and adipose tissue. Glucose is therefore preserved for use by the brain, which needs it most.

- The metabolic rate increases. Oxygen consumption and body heat rise.

- In the lungs, small tubules called bronchioles dilate, increasing the flow of air.

- Smooth muscles in the gastrointestinal tract and sphincters contract, while muscles in the uterus and trachea relax.

- Motor activity, gastrointestinal secretion, and other nonessential activities slow to conserve energy for other, more crucial functions.

Dopamine, a neurotransmitter, is similar to epinephrine. It influences the brain processes controlling emotion, movement, and the sensations of pleasure and pain. When dopamine is not produced in large enough quantities (for example, in patients with Parkinson's disease), the body grows rigid, and movement becomes difficult.

Adrenal Cortex

The large outer region of the adrenal gland, the cortex is made up of three layers or zones, each of which produces its own group of steroid hormones:

- **Zona glomerulosa**, the outermost layer, is where the mineralocorticoids (aldosterone) are produced.

- **Zona fasciculata**, the middle layer, is where the glucocorticoids (cortisol) are produced.

- **Zona reticularis**, the innermost layer, is where the gonadocorticoids (sex hormones androgens and estrogens) are produced.

The adrenal cortex is separated into two functional regions, each regulated by a separate entity. The fasciculata and reticularis layers depend upon adrenocorticotropic hormone (ACTH) stimulation from the anterior pituitary. Without ACTH, these two regions would atrophy. The zona glomerulosa is under the control of the renin-angiotensin system, which regulates blood pressure.

Mineralocorticoids, glucocorticoids, and gonadocorticods are all steroid hormones derived from cholesterol. Most of the cholesterol that enters the adrenal gland comes from **low-density lipoproteins** (LDLs) circulating in the blood. Upon stimulation of the adrenal cortex, cholesterol that splits off from the LDL is converted into pregnenolone, the precursor molecule for steroid hormones. The adrenal gland can also synthesize a small amount of its own cholesterol.

Mineralocorticoids

The principal mineralocorticoid, aldosterone, is produced and secreted by the zona glomerulosa. Aldosterone acts upon the kidneys to regulate sodium, potassium, and water reabsorption. It stimulates the distal tubules to reabsorb more sodium and excrete more potassium (as well as hydrogen) in the urine. As sodium is reabsorbed, more water is also reabsorbed, increasing blood fluid volume. Aldosterone also acts upon the salivary glands, sweat glands, and colon to reduce the amount of sodium lost in saliva, sweat, and feces.

Sodium and potassium balance is crucial, because these fluids regulate fluid movement between the cells and the extracellular fluid. Without aldosterone, a deadly fluid imbalance could result. When too much sodium is present, water from inside the cells crosses over into the extracellular region to restore balance, causing the cells to shrink—a situation that could lead to shock and death. Too little sodium can send water into the cells, causing them to swell and potentially leading to nausea, vomiting, diarrhea, convulsions, or coma.

Aldosterone release is primarily under the control of the renin-angiotensin system, which helps regulate blood volume and blood pressure. Inside the blood vessels of the kidneys are tiny receptors that can detect changes in blood pressure and extracellular fluid volume. When these sensors notice a drop in pressure and volume, they release the enzyme renin into the blood. Renin travels to the liver, where it converts the protein angiotensinogen into another protein, called angiotensin I. Once angiotensin I reaches the lungs, it is converted by angiotensin-converting enzyme (ACE) into the much more potent hormone, angiotensin II, which constricts the blood vessels and stimulates the zona glomerulosa to synthesize aldosterone, raising blood pressure.

Glucocorticoids (Corticosteroids): Cortisol (Hydrocortisone)

The principal glucocorticoid, cortisol, is produced in the zona fasciculata and is sometimes referred to as the "stress hormone." Like the catecholamines from the adrenal medulla, it helps the body respond during times of stress (for example, injury or emotional trauma). Cortisol is essential because of its effects on metabolism: It maintains the body's energy (glucose) supply and regulates fluid balance. Without it, the body could overreact to stress and disrupt the fragile homeostatic balance that it needs to stay alive.

The testes, erythrocytes, kidney medulla, and especially the brain rely on glucose as their sole energy source. Without it, they cannot function properly. After a meal, blood levels of glucose are typically high and the body is able to store whatever it does not immediately use. But glucose stores are not everlasting. Glycogen, the form of glucose stored in the liver, can run out within 24 hours after a meal. During periods of fasting, cortisol maintains blood glucose levels by affecting a number of metabolic processes.

Glucose, Fat, and Protein Metabolism

In the muscles and other tissues, cortisol increases the breakdown of protein into amino acids. Those amino acids are used to produce additional glucose (via a metabolic pathway called gluconeogenesis) in the liver. Cortisol also conserves glucose for the brain and spinal cord by blocking the actions of insulin (which will be discussed later in this chapter)—inhibiting glucose absorption into other tissues.

Cortisol also stimulates the release of fatty acids and glycerol from adipose tissue. Glycerol is used in gluconeogenesis, while fatty acids are made available for energy to other tissues to preserve glucose for the brain.

Cortisol reduces protein reserves everywhere except in the liver. As proteins continue to be broken down in muscles and in other tissues, blood levels of amino acids rise. The additional amino acids are used for gluconeogenesis, glycogen formation, and protein synthesis in the liver.

The Pancreas

The pancreas has two roles: It functions both as an endocrine and as an exocrine organ. As an exocrine organ, the pancreas releases digestive enzymes via a small duct into the small intestine. These enzymes break down carbohydrates, fats, and proteins from food that has been partially digested by the stomach. The exocrine pancreas also releases a bicarbonate to neutralize stomach acid in the duodenum (first portion) of the small intestine.

In its role as an endocrine organ, the pancreas secretes the hormones insulin and glucagon, which help the body use and store its primary source of energy—glucose (sugar). The endocrine pancreas also secretes somato-statin, which is a primary regulator of insulin and glucagon release.

The pancreas is long and soft, and stretches from the duodenum of the small intestine almost to the spleen. It is divided into a head (its widest point), neck, body, and tail. The endocrine pancreas is made up of clusters of cells called the **islets of Langerhans**, in which the hormones insulin and glucagon are produced. There are about one million islets, but they make up only about 1 percent of the endocrine pancreas' total volume. The islets also contain parasympathetic and sympathetic neurons, which influence insulin and glucagon secretion. The islets also produce the hormones somatostatin and **pancreatic polypeptide**.

Energy Metabolism

Why are insulin and glucagon so crucial? Because the body needs energy to survive, and these two hormones regulate the distribution of energy to tissues. Energy enters the body in the form of food. As food passes through the mouth, esophagus, and stomach, enzymes break it down into tiny pieces. Once the partially digested food reaches the intestines, more enzymes go to work, breaking it down into molecules small enough to enter the bloodstream and be transported to cells. Starches are broken down into glucose (sugar), proteins are broken down into amino acids, and fats are broken down into fatty acids and glycerol.

Food metabolism occurs in two distinct phases: During the anabolic phase, which occurs after a meal, enzymes convert nutrients from food into substances the body can use. Blood levels of glucose, fatty acids, and amino acids rise. Because the body has more energy than it needs at the moment, it stores the excess for later. Glucose is stored as glycogen in the liver and muscles, fat is stored in adipose tissue, and amino acids are stored in muscle.

About four to six hours after a meal, the catabolic phase begins. Stored energy from the liver, muscles, and adipose tissue is mobilized to sustain the body until its next meal. The liver produces glucose from stored glycogen and by converting amino acids via gluconeogenesis. When the body has gone for some time without food, the liver converts free fatty acids into ketone bodies. The brain normally uses only glucose for energy, but it can use ketone bodies as a backup energy source when glucose supplies are low. Without this alternative energy source, the brain and nervous system would starve and suffer permanent damage.

The hormones insulin and glucagon from the endocrine pancreas regulate these stages of energy metabolism. Insulin primarily regulates the anabolic phase, while glucagon influences the catabolic stage.

Insulin

After a meal, the body converts carbohydrates from foods into simple sugars in the intestine. Glucose is carried to the tissues through the bloodstream. When blood glucose levels rise, the beta cells in the endocrine pancreas produce and release insulin. Insulin is formed from a larger, inactive molecule, called **proinsulin**. Before insulin is released into the bloodstream, the inactive molecule splits off. For a look at diabetes, a chronic condition involving insulin production, see Sidebar 4.1.

Insulin levels rise 8–10 minutes after a meal, reaching their peak concentration 30–45 minutes after the meal. Nearly every cell in the body has insulin receptors. When insulin binds to its receptors on the cell surface, it triggers other receptors that help the cells take in glucose. The body uses and stores glucose in the liver, muscles, and adipose tissues. Without insulin, an individual could eat three meals a day and still starve to death because the cells would be unable to use the energy.

In the liver, insulin promotes glucose storage in the form of glycogen. It also inhibits the breakdown of glycogen and the production of glucose from other, noncarbohydrate sources (gluconeogenesis), and it decreases overall glucose release by the liver.

Insulin helps transport glucose to muscle cells and stimulates the incorporation of amino acids into protein, which is used to sustain and repair muscles. It promotes glycogen synthesis to replace glucose the muscles have used. Insulin also promotes glucose uptake in adipose tissue, promotes its conversion to fatty acids, and inhibits the release of stored fatty acids.

As insulin moves glucose into the tissues for energy use and storage, blood glucose levels fall. Between 90 and 120 minutes after a meal, blood glucose concentration returns to its original, pre-meal levels. To help the

SIDEBAR 4.1

Understanding Diabetes: The Key to Staying Healthy

Diabetes is a serious, chronic condition involving the endocrine system. It is defined as a disease where patients have elevated levels of blood glucose or sugar that results from defects in either or both insulin production and action. According to the most recent estimates from the National Institute of Diabetes and Digestive and Kidney Diseases (NIDDK), 23.6 million people—over 7 percent of the population—have diabetes. Of particular concern is that of these over 20 million people with diabetes, almost 6 million are undiagnosed. This is especially dangerous because diabetes can lead to serious complication—even premature death—if patients do not control the disease and lower the risk of complications.

There are three primary types of diabetes, and below is a brief explanation of each of these types:

- *Type 1 Diabetes:* Formerly known as insulin-dependent diabetes mellitus or juvenile-onset diabetes, type 1 diabetes occurs when the body's immune system destroys the cells in the body—called pancreatic beta cells—that produce insulin that regulates blood glucose levels. Patients with this type of diabetes use either a pump or injection device to deliver insulin into their body. While type 1 diabetes typically develops in children and young adults, it can develop in people of any age. In fact, type 1 diabetes accounts for 5–10 percent of all diagnosed cases of diabetes. There is currently no known way to prevent this type the diabetes. Some risk factors are believed to be linked to autoimmune, genetic, or environmental factors.

- *Type 2 Diabetes:* Formerly known as non-insulin-dependent diabetes mellitus, this type of the disease is estimated to comprise 90–95 percent of all diagnosed cases of diabetes in adults. Diagnosis is typically preceded by an inability of the cells to properly produce insulin—a condition known as insulin resistance. The need for insulin steadily increases while the pancreas generally loses the ability to produce it.

  This type of diabetes has several known risk factors: older age, obesity, family history of the disease, and lack of physical activity. In addition, experts believe that race/ethnicity can be a risk factor. Medical experts believe that African Americans, Hispanic/Latino Americans, Native Americans, and some Asian Americans and Native Hawaiians or other Pacific Islanders have an increased risk for type 2 diabetes.

- *Gestational Diabetes*: This type of diabetes is related to glucose intolerance and is diagnosed in pregnant women—occuring more frequently among African American, Hispanic/Latino American, and Native American women. Obesity and family history are also risk factors for diabetes. During pregnancy, treatment is focused on normalizing glucose levels to protect the infant. After pregnancy, however, an estimated 5–10 percent of women with gestational diabetes are diagnosed with type 2 diabetes. In addition, women with gestational diabetes are at an increased chance—40–60 percent—of developing type 2 diabetes over the next 5–10 years.

If not treated and controlled properly, diabetes can lead to serious complications, including blindness, kidney damage, cardiovascular disease, and amputations of lower limbs. The risk of these complications can be lowered by controlling blood glucose, blood pressure, and blood lipid levels. As mentioned earlier, many type 1 and 2 diabetes patients need insulin delivered by injection or a pump. Some patients also have to take other medications to control related conditions, such as high blood pressure. However, some patients take little or no medications—even insulin—by following a healthy meal and exercise plan. According to the NIDDK, among adults with type 1 or 2 diabetes, an estimated 14 percent take only insulin, 13 percent take both insulin and oral medication, 57 percent take oral medication only, and 16 percent do not take either insulin or oral medication. The treatment regimen for each individual will often change over the course of the disease. Among diagnosed diabetics—type 1 or type 2—14 percent take insulin only, 13 percent take both insulin and oral medication, 57 percent take oral medication only, and 16 percent do not take either insulin or oral medication. Medications for each individual with diabetes will often change over the course of the disease.

body maintain a constant blood glucose level, insulin and glucagon release are synchronized on an alternate schedule. When glucose concentrations in the blood rise during the anabolic phase, insulin is released. As insulin pulls glucose from the blood for tissue use and storage, blood glucose concentrations drop, stimulating glucagon release.

Insulin release may also be triggered by signals from the nervous system in response to external stimulation; for example, the sight and/or smell of food. The gastrointestinal hormones cholecystokinin (CCK), secretin, gastrin, **gastric inhibitory peptide** (GIP), and acetylcholine are thought to play a role in this response, preparing the pancreas to release insulin. Insulin release is inhibited not only by low glucose levels, but also by low levels of amino acids and fatty acids in the blood, as well as by the hormones somatostatin, epinephrine, and **leptin**.

Glucagon

Following a meal, insulin pulls glucose from the blood to be used and stored by cells. When several hours have passed without additional food being ingested, blood sugar is eventually depleted (a condition called hypoglycemia). The body still needs energy, much of which it gets from fatty acids until the next meal is available. But the brain, which cannot directly use fatty acids and other alternative energy sources, still relies on glucose. In response to dropping blood glucose levels, the alpha (α) cells of the endocrine pancreas begin to secrete the hormone glucagon from the large precursor molecule **proglucagon**. The same prohormone is found in cells of the gastrointestinal system, although it produces different secreted products.

Glucagon has the opposite effect of insulin. Whereas insulin lowers blood glucose levels by promoting glucose usage and storage, glucagon raises blood glucose levels. It acts primarily upon the liver to increase glucose output. When it binds to receptors on liver cells, glucagon activates the enzymes that break down stored glycogen (glycogenolysis) to release glucose and increases production of glucose from amino acid precursors (a process called gluconeogenesis). In adipose tissue, glucagon promotes the breakdown and release of fatty acids (lipolysis) into the blood, which are used by the cells for energy in the absence of glucose. By raising the level of fatty acids in the blood, glucagon indirectly prevents glucose uptake by the muscles and adipose tissue.

The main trigger for glucagon release is low blood sugar, but it may also be stimulated by other hormones, namely the catecholamines (in stressful situations); cholecystokinin, gastrin, and gastric inhibitory peptide (GIP) from the gastrointestinal system; and the glucocorticoids. Sympathetic nerve stimulation can also lead to glucagon release. Rising blood glucose levels, high circulating levels of fatty acids, as well as the hormones insulin and somatostatin inhibit glucagon secretion.

Somatostatin

The hormone somatostatin is produced in the delta (δ) cells of the pancreatic islets as well as in the gastrointestinal tract and hypothalamus. Somatostatin is primarily an inhibitory agent. In the pancreas, it acts in a paracrine manner, suppressing production of insulin and glucagon. It also acts upon the gastrointestinal tract, inhibiting secretion of the hormones gastrin, secretin, and cholecystokinin; prolonging gastric emptying time; decreasing gastric acid and gastrin production; and slowing intestinal motility. Together, these actions reduce the rate of nutrient absorption. When secreted from the hypothalamus, somatostatin acts upon the pituitary to inhibit growth hormone secretion.

Somatostatin release is triggered by rising levels of glucose, fatty acids, and amino acids in the blood. Gastrointestinal hormones like secretin and cholecystokinin can also stimulate its release. Insulin inhibits somatostatin secretion.

Gastrointestinal Hormones

The intestinal tract not only digests food and absorbs nutrients; it also produces and secretes a number of hormones that aid in the digestive process. Gastrointestinal hormones, which are primarily peptides, are produced in specialized endocrine cells in the stomach and small intestine, as well as in neurons scattered throughout the gastrointestinal tract. Most gastrointestinal hormones either act upon nearby cells (paracrine delivery) or act as neurotransmitters within neurons (neurocrine delivery). Endocrine and neural cells in the intestinal tract are referred to collectively as the enteric endocrine system.

The central nervous system and gastrointestinal tract are linked by pathways known as the brain-gut axis. Neurotransmitters located in both the brain and the gut regulate and coordinate such functions as satiety, nutrient absorption, gut motility, and intestinal blood flow. Any disruption of this system is believed to result in gastrointestinal disorders such as irritable bowel syndrome (IBS).

Gastrin

This hormone, produced by specialized cells in the stomach, regulates stomach acid secretion. When partially digested proteins, peptides, and amino acids are present in the stomach, gastrin stimulates the release of gastric acid and the digestive enzyme, pepsin, into the stomach cavity to aid digestion. Beer, wine, and coffee can also stimulate gastrin release. As the stomach becomes more acidic, gastrin production declines.

Cholecrystokinin (CCK)

Food entering the small intestine must be broken down into smaller molecules (such as amino acids and fatty acids) in order to be absorbed. When partially digested proteins and fats enter the duodenum (first portion) of the intestine, cells in that region secrete the peptide hormone CCK. This hormone triggers the release of pancreatic enzymes and stimulates gallbladder contractions to release stored bile. Pancreatic enzymes and bile are sent to the intestines, where they aid in digestion. As proteins and fats are digested and absorbed, a drop in their levels shuts off CCK release. CCK is also released as a neurotransmitter by the central nervous system. Scientists believe that it may help regulate food intake by signaling the feeling of satiety.

Secretin

The stomach regularly secretes acid, which could potentially burn and damage the small intestine. When acid enters the duodenum, cells lining the region release secretin. This hormone stimulates the pancreas to release acid-neutralizing bicarbonate and water. As acid in the intestine is neutralized, the rising pH level shuts off secretin release. Somatostatin also inhibits secretin production.

Gastric Inhibitory Peptide (GIP)

Gastric inhibitory peptide (GIP), also called glucose-dependent insulino-tropic peptide, is part of the secretin family of hormones, and like secretin, it is released by cells in the duodenum. It blocks gastrin and gastric acid secretion into the stomach and inhibits gut motility. Following a meal, GIP responds to increased glucose levels, enhancing the insulin response to glucose.

Vasoactive Intestinal Polypeptide

This peptide is found throughout the body but is secreted in greatest concentration by cells in the intestinal tract and nervous system. VIP increases secretion of water and electrolytes by the intestine, increases blood flow within the gut, and inhibits gastric acid secretion.

Ghrelin

This peptide hormone, secreted by epithelial cells in the stomach, stimulates growth hormone secretion from the pituitary gland. In the gastrointestinal system, ghrelin stimulates the sensation of hunger by communicating the body's energy needs to the brain. Ghrelin levels in the blood are highest during the fasting state several hours after a meal, and lowest just after food consumption. Scientists are investigating ghrelin's role in obesity, in the hope of one day discovering more effective weight control methods.

Motilin

As its name suggests, **motilin** controls movement (smooth muscle contractions) in the gut. In between meals, cells in the duodenum secrete small bursts of motilin into the blood at regular intervals. Motilin contracts and releases smooth muscles in the intestine wall to clean undigested materials from the intestine.

Substance P

The neuropeptide **substance P**, found in both the brain and gastrointestinal system, stimulates smooth muscle contractions and epithelial cell growth. It may also be involved in inflammatory conditions of the gut. In the brain, substance P has been linked to both the pain and pleasure responses. It is released from enteric neurons in response to central nervous system stimulation, serotonin, and CCK, and it is inhibited by somatostatin.

Endocrine Functions of the Sex Glands

Sex Glands

The sex glands (ovaries in the female and testes in the male) serve as both reproductive and endocrine organs. They produce the eggs and sperm that form the basis of human life. They also synthesize and secrete the sex steroids—testosterone, estrogen, and progesterone. These hormones give males and females their individual sexual characteristics, and play a key role in reproduction.

The two bean-shaped ovaries (Figure 4.10) sit on either side of the female uterus, just below the openings to the fallopian tubes. Like the adrenal glands, the ovaries contain an outer cortex and an inner medulla. The medulla consists primarily of connective tissue containing blood vessels, smooth muscle, and nerves. The real activity occurs in the larger outer cortex, which holds the follicles in which the eggs develop. Eggs are stored inside these follicles until they are ready to be released on their journey through the fallopian tubes, where they may ultimately be fertilized by sperm.

Also inside the cortex are specialized cells that produce and secrete the steroid hormones estrogen and progesterone, as well as less potent male hormones (androgens). Ovarian sex hormones are produced and released in response to follicle-stimulating hormone (FSH) and luteinizing hormone (LH) from the anterior pituitary. Once released, estrogens and progestins influence the development of the female reproductive organs and sexual characteristics. On the side of each ovary is a small notch, called the hilum, through which blood vessels and nerves enter and exit.

The ovaries and testes are necessary for reproduction: Without them, the human species could no longer reproduce. Along with their central

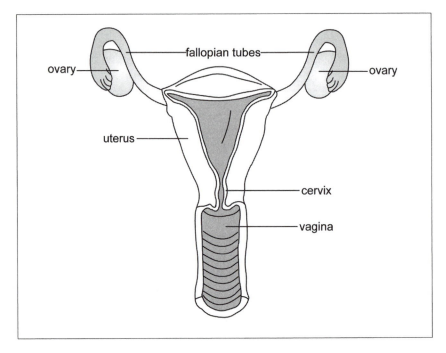

Figure 4.10   The ovaries. (Sandy Windelspecht/Ricochet Productions)

functions—producing and sustaining the eggs and sperm—the sex glands have a number of other crucial tasks. As part of the endocrine system, they produce and secrete the sex hormones, which are involved in sexual maturation, differentiation, and function, as well as metabolism and bone growth.

The Ovaries

The ovaries are the female reproductive organs. Inside the ovaries, the eggs, or ova, develop and are nourished until they are ready to be released into the fallopian tubes. The eggs form and develop by a process known as **oogenesis**. In the developing female embryo are primordial germ cells, from which the fundamental egg cells develop. These cells remain dormant until a girl reaches puberty, then develop into the mature eggs, which are released one at a time in a process known as ovulation. Like the adrenal gland, each ovary is constructed of an outer cortex and an inner medulla. The cortex contains tiny sacs called follicles. Ovarian follicles consist of two types of cells: theca and granulosa. Nestled inside each follicle is an immature egg, surrounded by a layer of granulosa cells. A woman is born with every egg (about a million of them) that she will ever possess. Once these eggs are used up, she will no longer be able to conceive.

The ovarian follicles normally remain in an inactive state. These so-called primordial follicles lie in wait for hormonal stimulation that will help them mature and prepare them for possible fertilization. During each menstrual cycle, follicle-stimulating hormone (FSH) from the pituitary stimulates a few of these eggs. Typically, only one egg completes the ovulation process.

Follicular Phase

During this phase, which lasts 10–16 days, gonadotropin-releasing hormone (GnRH) from the hypothalamus triggers the release of follicle-stimulating hormone (FSH) and luteinizing hormone (LH) from the anterior pituitary. FSH and estrogen stimulate the development of between 6 and 12 primary follicles. LH causes the follicles to produce estrogen. Between days 5 and 7 of the cycle, one of the follicles ripens (in the case of a multiple pregnancy, more than one follicle has ripened) and becomes ready for ovulation. As the dominant follicle develops, estrogen levels rise, inhibiting FSH secretion. Without FSH stimulation, the other follicles begin to wither away.

Ovulatory Phase

At the end of the follicular cycle, estrogen levels peak. Estrogen normally suppresses gonadotropin production, but in this case, rising estrogen levels trigger a surge in LH and FSH (usually around day 13 of the cycle). The hormonal surge lasts between 24 and 36 hours, at the end of which the dominant follicle ruptures and releases its egg from the ovary. This is called ovulation.

Luteal Phase

Following ovulation, the erupted follicle transforms into a body called the corpus luteum. The corpus luteum begins to secrete estrogen and progesterone under the influence of luteinizing hormone from the pituitary. Estrogen and progesterone prepare the uterus for implantation and are necessary to maintain a pregnancy. If the egg is fertilized, the corpus luteum remains intact and continues to secrete estrogen and progesterone throughout the first trimester of pregnancy. The luteal phase lasts for about 14 days.

Menstrual Phase

If the egg is not fertilized, production of progesterone and estrogen in the corpus luteum diminishes. The uterine lining, which has become rich with blood vessels to nourish the growing embryo, is no longer needed. Small arteries in the lining constrict, cutting off oxygen and nutrients. The cells die and slough off (known as menstruation). The corpus luteum undergoes a process known as **luteolysis**. It degenerates, becomes unable to produce hormones, and is finally replaced by scar tissue. As steroid production by the corpus luteum decreases, FSH secretion increases, stimulating the development of new follicles and initiating a new menstrual cycle. Menstruation lasts for four to five days.

In addition to providing a site for egg development, the ovaries produce several steroid and peptide hormones. Steroid hormones (estrogens, progesterone, and androgens) are produced in the follicular cells. Like all steroid hormones, they are produced from cholesterol, which is both present in the ovaries and transported to the ovaries in the form of low-density lipoproteins (LDLs). Peptide hormones (relaxin, inhibin, oxytocin, and vasopressin) are produced in the follicular cells and within the corpus luteum.

Estrogen

The primary female sex hormones produced in the ovaries, estrogens play a role in the development of sexual characteristics and help regulate the reproductive cycle. The three estrogens—estradiol, estriol, and estrone—are produced in the thecal and granulosa cells in the developing follicles.

Estradiol is the most powerful—and most plentiful—of the three estrogens. It is made from the androgens testosterone and androstenedione, which are produced in the thecal cells under the influence of luteinizing hormone. In the granulosa cells, follicle-stimulating hormone helps convert these androgens into estradiol (and estrone). At the onset of puberty, estradiol influences maturation of the reproductive organs (uterus, fallopian tubes, cervix, and vagina) and redistributes fat to the hips, buttocks, thighs, and breasts to produce a more feminine, curvy shape.

Estradiol also influences the menstrual cycle by stimulating and inhibiting the release of LH and FSH. Normally, estradiol acts upon the hypothalamus to inhibit GnRH secretion, which prevents the release of LH from the anterior pituitary. But at the end of the follicular phase of the menstrual cycle, rising estradiol concentrations trigger the surge of LH that initiates ovulation.

Most of the estriol in a woman's body is produced not in her ovaries but in her liver, where it is converted from estrone and (by a more indirect route) estradiol. This relatively weak hormone may actually act as a partial agonist–partial antagonist by blocking receptors that would otherwise be occupied by the stronger estrogen, estradiol. During pregnancy, estriol is secreted in large quantities by the placenta. Doctors test a mother's urine for this hormone to assess the viability of her pregnancy.

Estrone, the weakest of the three estrogens, is primarily converted from estradiol or androstenedione (from the adrenal cortex). It is similar in function, although not as potent, as estradiol. Following menopause, estrone production increases due to increased conversion of androstenedione.

When the estrogens are secreted into the bloodstream, they travel bound to proteins—mainly albumin and sex hormone-binding globulin (SHBG). Once they arrive at their target cells, estrogens bind to an intracellular protein, which carries them to the nucleus. There, they influence protein synthesis.

Progestins

Progestins are primarily designed to maintain and support a pregnancy. Synthetic versions can also prevent a pregnancy. The most significant progestin is progesterone. It is produced by the corpus luteum after the egg has been released from the follicle. If the egg is fertilized, the corpus luteum continues to produce progesterone for the first trimester of pregnancy until the placenta takes over production. Progesterone prepares the mother's body for pregnancy by thickening the uterine lining to nourish the growing embryo. It then maintains the viability of the pregnancy by stopping additional follicles from becoming mature and by preventing uterine contractions.

Progesterone travels through the blood bound to corticosteroid-binding globulin (CBG) and albumin. Its other actions are to stimulate breast growth and development (along with estrogen); influence carbohydrate, protein, and fat metabolism; and decrease the body's responsiveness to insulin (as sometimes occurs during pregnancy).

The Testes

The testes, like the female ovaries, serve both reproductive and endocrine functions (Figure 4.11). They are made up of a network of tubules

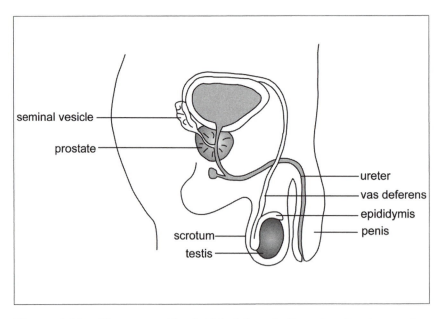

Figure 4.11 The testes. (Sandy Windelspecht/Ricochet Productions)

(**seminiferous tubules**) that produce and carry sperm, interspersed with cells in which androgens (male hormones) are produced. The testes contain two types of cells:

Sertoli cells: These cells, which line the seminiferous tubules, surround and nourish the germ cells from which sperm develop, and they facilitate the journey of sperm out of the testes. Sertoli cells also produce androgen-binding protein, which maintains high levels of androgens in the testes and seminal fluids, as well as many peptides (inhibin, activin, and follistatin) that regulate testicular function.

Leydig cells: In between the seminiferous tubules are the cells in which testosterone, the primary androgen, is produced.

Sperm Production

Production of sperm in the Sertoli cells depends upon stimulation by follicle-stimulating hormone (FSH) from the anterior pituitary and by testosterone. When FSH binds to androgen receptors on Sertoli cells, it stimulates the production of androgen-binding protein. This protein keeps levels of testosterone in the seminiferous tubules high. FSH stimulates sperm production and maturation. As androgen levels rise, Sertoli cells begin to secrete inhibin, which inhibits FSH release from the pituitary.

Testicular Hormones

The testes produce a number of hormones, including testosterone, dihydrotestosterone, androstenedione, and estradiol. Testosterone is by far the most plentiful, and most important, of these hormones.

Like other steroid hormones, testosterone is synthesized from cholesterol. Its release is initiated by gonadotropin-releasing hormone (GnRH) from the hypothalamus, which stimulates the release of luteinizing hormone (LH) from the pituitary. LH stimulates the Leydig cells to produce and secrete testosterone. In a classic negative-feedback loop, elevated testosterone levels in the blood inhibit secretion of LH by acting on the hypothalamus and the anterior pituitary, both of which contain androgen receptors.

Testosterone travels through the blood bound to carrier proteins—typically either sex hormone–binding globulin (SHBG) or albumin. When it reaches its target cell, it binds to the androgen receptor.

Summary

The human body must have certain communications in place that coordinate the function of all of its organs, muscles, nerves, and tissues. There are actually two communication centers in the human body—the nervous system and the endocrine system. While the nervous system communicates through electrical impulses traveling throughout the body as part of a nerve network, the endocrine system communicates through chemicals called hormones, which travel through the bloodstream. In fact, hormones actually control how the nervous system behaves through stimulation or inhibition. Hormones are secreted through the glands of the endocrine system. The glands are the hypothalamus, pituitary, thyroid, parathyroids, adrenal, pancreas, and sex glands.

5

The Integumentary System

Julie McDowell

Interesting Facts

- The entire body is covered with skin. This surface area is between 1.5 m^2 and 2 m^2.

- Skin makes up approximately 7 percent of the body's weight. It weighs approximately 4 kg.

- It's estimated that approximately 70 percent of household dust is made up of shed human skin.

- Between 30,000 and 40,000 dead skin cells drop off the body every minute.

- Nails grow an average of two centimeters every year, and fingernails grow almost four times as fast as toenails.

- There are approximately 10,000 hairs on a human head, with each hair growing a rate of five inches a year.

- Human hair is virtually impossible to destroy; it is resistant to extreme cold and heat (except for burning), water, and many types of acids and chemicals.

- Over the course of a lifetime, each human will shed an average of 40 pounds of skin.

- On average, each person loses between 80 and 100 hairs every day.

- Each human scalp has an average of 100,000 hairs.

Chapter Highlights

- Skin: epidermis, dermis

- Skin's accessory organs: hair and hair follicles, nail follicles

Words to Watch For

Blister	Hair	Phagocytize
Callus	Hair root	Pore
Cerumne	Hair shaft	Sebaceous duct
Ceruminous glands	Hyponchium	Sebaceous glands
Collagen	Keratin	Skin
Elastin	Langerhans cells	Skin grafts
Encapsulated nerve endings	Lunula	Stratum corneum
	Melanin	Stratum
Eponychium	Melanoctyes	germinativum
Free nerve endings	Papillary layer	Sweat glands

Introduction

The integumentary system is made up of **skin** and the subcutaneous tissue, which is located right below the skin's surface (Figure 5.1). This system also includes the skin's two accessory structures, the **hair** and **sweat glands**, as well as glands and sensory receptors. Because it covers the entire surface of the body, it is an important protector of the body by keeping out substances that could be harmful. In fact, the skin is actually the body's largest organ, and has many important roles, including helping to regulate the body's temperature. (For more details on some of the jobs that the skin and subcutaneous tissue performs for the body, see Sidebar 5.1. Some of the jobs will be discussed in more detail throughout this chapter, as well as this encyclopedia.)

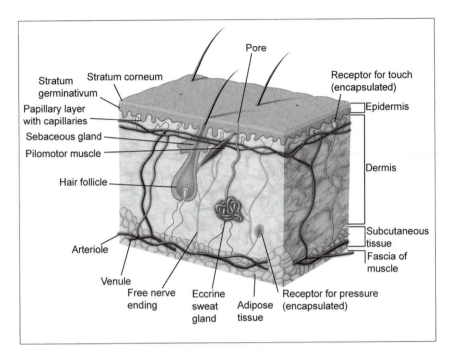

Figure 5.1   The Body's Skin Structure. The body's subcutaneous tissue. (Andreus/Dreamstime.com)

Skin

Epidermis

The skin has two primary layers: the epidermis (outer) and the dermis (inner). The epidermis has two important sublayers—the stratum germinativum and the stratum corneum. The epidermis is made up of cells called keratinocytes; these cells make up the epithelial tissue.

It is in the epidermis layer where **blisters** and **calluses** can form. Blisters occur when friction—such as what might develop between the skin and the inside of a shoe—causes layers to separate within the epidermis or between the epidermis and the dermis. As these layers separate, tissue fluid may build up, leading to a blister. While blisters often result because of friction, calluses result from pressure. When exposed to increased amount of pressure, mitosis will occur at a rapid rate, causing the epidermis to thicken.

SIDEBAR 5.1

Skin: The Body's Best Protector, and More

In addition to protecting the body from dangers related to the environment, the skin also has a number of other important jobs:

- *Regulating Body Temperature*: The blood vessels in the dermis layer of the skin widen (vasodilate) and narrow (vasoconstric), allowing the production of sweat that cools the body.

- *Sensing Pain*: There is a thick collection of nerve endings located in the dermis that is sensitive to pain as well as pressure. These nerves communicate with the brain through the nervous system, informing the brain when danger is imminent, such as in the form of a hot surface (heat).

- *Losing Too Much Water*: Within the dermis layer, glands secrete a substance known as sebum. This spreads across the skin, enabling it to become waterproof. The dermis also has collagen fibers that soak up water.

- *UV Protection*: The skin produces a pigment called melanin, which filters the sun's dangerous ultraviolet (UV) radiation.

- *Manufacture of Vitamin D*: Produced by the skin as a result of exposure to sunlight, vitamin D plays an important role in regulating how the body metabolizes calcium.

Calluses can occur anywhere on the skin, although they are most often found on the palms of the hands and soles of the feet.

Of the two epidermal sublayers, the innermost layer is the **stratum germinativum**. This is the surface where mitosis occurs, meaning that new cells are always being produced while older cells are forced to the skin's surface. These new cells are producing a protein called **keratin**. As the older cells are pushed farther away from the dermis, they eventually die. These dead skin cells eventually make their way to the outer sublayer, the **stratum corneum**.

While the cells are dead by the time they get to the outer layer of the skin, they do still contain keratin. This protein is important to the skin— it is waterproof, prevents evaporation of water inside the body, and prevents water from entering the body through the skin. If the body did not have a waterproof statum corneum layer, we would not be able to take a bath or shower, or even swim without risking serious damage to our bodies.

In addition to its waterproofing functions, the stratum corneum also protects the body from dangerous substances such as pathogens, like bacteria, and chemicals that could be dangerous. Of course, if skin is broken, then some of these substances can enter the body, which could lead to an infection. The body is even more vulnerable if it is burned, especially if the burn is so bad that the stratum corneum is destroyed (see Sidebar 5.2 for more information on the impact that burns can have on the body).

The epidermis also contains two types of cells that are important protectors: the **Langerhans cells** and the **melanoctyes**.

Langerhans cells originate in the bone marrow and are known for their ability to travel fast within the body. They need to be mobile, because they are an important part of the body's immunity response. When a bacteria or other foreign substance enters the body through broken skin, the Langerhans cells ingest or **phagocytize** the pathogen, and then deliver it to the lymph nodes. There, a type of white blood cell called a lymphocyte prompts the start of the body's immune response and antibodies are produced to neutralize any danger presented to the body by the pathogen. (For more detail on the body's immune response, see Chapter 6 on the lymphatic system.)

The second type of important cell in the epidermis, the melanocytes, produces a kind of protein called **melanin** (also known as pigment) that plays an important role in skin color. Skin color is genetic or hereditary, and is determined by how fast and how much melanin is produced by the body's melanocytes. In dark-skinned people, their bodies are constantly producing a large amount of melanocytes, while light-skinned people's bodies produce much less of this protein. But regardless of one's skin color, exposure to the sun's ultraviolet (UV) rays causes melanin production to ramp up. Change in a skin's color—sunburn or tanning—is the result of excess melanin making its way from the epidermal cells to the skin's surface. Another result of prolonged sun exposure is freckles, which

---

SIDEBAR 5.2

Burns

The skin can be burned by sources of heat, such as fire, steam, hot water, or sunlight. But it can also be burned by electricity or certain chemicals that are corrosive. Burn injuries can be minor or fatal. They are classified in three categories, according to the resulting damage:

- *First-degree Burns*: While a first-degree burn can be painful (think of a bad sunburn), it is minor because only the superficial or outermost layer of the epidermis is affected. First-degree burns do not result in blisters on the skin's surface. The skin will appear red, indicating an inflammatory response. The inflammation means that vasodilation is occurring, bringing additional blood to the burn site.

- *Second-degree Burns*: Blisters occur as a result of second-degree burns, which affected the deeper layers of the epidermis. In a second-degree burn, the inflammation prompts the release of histamine, causing the blood capillaries to release plasma. This plasma then becomes tissue fluid, which form into blisters as it collects on the skin's surface.

- *Third-degree Burns*: These burns destroy the entire epidermis, including its outermost layer, the stratum corneum, that is a barrier from pathogens and other dangerous chemicals from entering the body. Ironically, because the epidermis is gone, so are the pain receptors, so victims of these burns often do not initially feel much pain. But because the protection of the stratum corneum is lost, the victims are at serious risk of developing infections, as well as dehydration, because the skin is not there to keep the water in the body.

---

represent concentrated areas of cells that produce melanin. The darkening of the skin actually protects the stratum germinativum from overexposure to UV rays. As a result, dark-skinned people have some natural defense against the sun's harmful effects, such as skin cancer, as opposed to light-skinned people.

Dermis

As mentioned earlier in this chapter, the epidermis is the skin's outer layer, while the dermis is the inner layer. While the epidermis is an important protector of the body from dangers of the outside environment, the dermis is notable for its strength and flexibility. The tissue of the dermis is fibrous. It is made of two fibers: **collagen** and **elastin**. The collagen fibers are strong, while the elastin fibers are flexible and can return to their original shape after being stretched. But this elasticity does not last forever. With age, the elastin fibers break down, meaning that this elasticity undergoes deterioration, which results in wrinkles.

Located between the epidermis and the dermis is a layer called the **papillary layer**, which contains a large population of capillaries. Since there are no capillaries in the epidermis, the papillary layer is the epidermis' only access to oxygen and nutrients from the body's blood supply. These capillaries also provide oxygen and other nutrients to the dermis.

Before describing the accessory organs of the skin, such as hair follicles and nails, it is important to understand the skin's ability to repair itself. If the skin is cut—whether from an injury or surgical incision—it will automatically grow back together, even without stitches. What is really happening, however, is new skin is formed or regenerated. At the site of the cut, the skin cells located right next to the wound site enlarge, while the cells surrounding the wound multiply at a rapid pace to replace cells that have been destroyed. Healing begins as all of these cells eventually converge and epithelial cells continue production. Eventually, skin will be replaced at its original thickness.

This repair process also allows **skin grafts**—or transplanting of the skin—to be successful. Skin grafts are often necessary when a person has third-degree burns, because the skin is so damaged that it cannot regenerate. The skin to be transplanted is typically taken from a fleshy place on the body, such as the thighs or buttocks. Once this transplanted skin is applied to the wound, the cell regeneration process hopefully will occur, helping to cover the wound.

The Skin's Accessory Organs

Located in the dermis are the skin's accessory organs. These include the hair and nail follicles, sensory receptors, and various glands.

Hair and Hair Follicles

Hair follicles are composed of epidermal tissue, but their base is the **hair root** (Figure 5.2). It is at the hair root where mitosis occurs, and these new cells produce keratin. Melanin is involved because it determines the hair color. But these cells quickly die, and then become part of the **hair shaft**. The end of the hair follicle is called a bulb. In this bulb is located a supply of capillaries that feed the needs of the growing hair shaft. Hair type is also determined by the shape of this hair shaft. If the cross section of the shaft is round, the hair is curly, and the rounder the shaft, the curlier the hair.

Along with the hair root, the follicle also contains certain glands and nerve endings, as well as muscle. When hair follicles are located on the skin's surface, such as on the scalp, they are adjacent to oil or **sebaceous glands**. These glands produce an oily liquid called **sebum** that travels through the hair shaft to the body's surface. Sebum is important because it moisturizes the skin to keep it from drying out. It also is composed of bacteria-fighting substances that protect the follicles from infection. Sebum travels to the hair shaft by draining through a **sebaceous duct** in the hair follicle. The volume of sebum that the sebaceous duct produces depends not only on the actual size of the duct, but also the amount of hormones circulating, particularly the male sex hormones called **androgens**. The largest sebaceous ducts are located on the head, neck, and front and back of the chest. (For information on one of the most common skin disorders of the sebaceous glands, acne, read Sidebar 5.4 at the end of this chapter).

In addition to the sebaceous glands, nerve endings are located around the hair follicle's bulb. These nerves are affected by any movement near the hair shaft, meaning that if there is any pressure on or around the hair shaft (such as pulling of the hair), then these nerve endings will send signals to the brain. One example is when there is a bee or other insect flying on or around the hair on the top of the head. The nerve endings in the hair shaft will let the brain know that this insect is near, causing the person to swat it away, before it lands on the head and possibly plants its stinger, in the case of a bee.

As noted earlier, there are also muscles in the hair follicle. This muscle is called the **arrector pili**, which means "raiser of the hair" and is attached to the follicle. The contraction of the arrector pili forces the hair to move

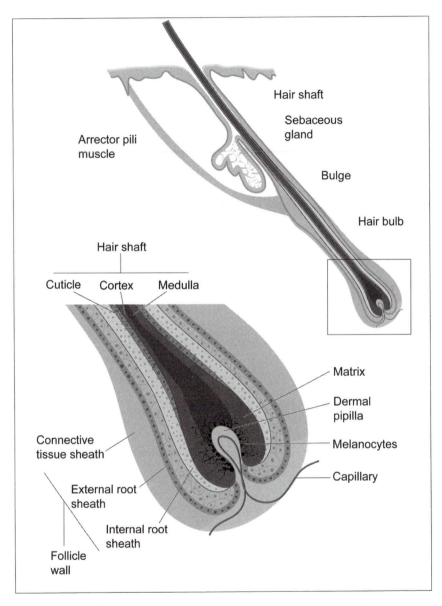

Figure 5.2   The Hair Follicle. Each hair root sits in a follicle, which is located about 4 or 5 mm below the skin surface. The sebaceous glands produce oil that lubricates the hair. (Legger/Dreamstime.com)

from its normal position—which is angled—and become vertically erect. In humans, the contraction of a significant number of these muscles occurs during times when the body is chilled or a person is afraid. It is also known as "goose pimples." This feature is an important source of protection for mammals with a lot of fur, because it allows them to insulate themselves by trapping air within their fur.

While hair is not as important to humans in comparison to other mammals, it does serve certain functions. It insulates the head and, as mentioned earlier, has nerves that sense when small objects or insects are close to the head.

There are millions of hair strands covering the body, but the most concentrated amounts of hair are on the head, surrounding the external genitalia, and under the armpits. The hair on our head—as well as our eyelashes and eyebrows—is actually these dead cells. Each strand contains the protein keratin, which is produced in the dermis. However, the actual strand is produced by a process called "inpouching," which occurs in the epidermis.

There are three layers to each hair strand: the medulla, cortex, and cuticle. The medulla is the central core, while the central cortex is the next layer. The cortex is made up of layers of flattened cells containing melanin that determines hair color. The outermost layer is the cuticle, which is made up of one layer of overlapping cells. This layer tends to break away at the end of hair strands, causing what is known as "split ends." (For information on hair thinning and loss, please see Sidebar 5.3.)

Nail Follicles

The primary function of nails is to protect the fingers and toes, and make it easier for the fingers to pick up small objects (Figure 5.3). While the nails themselves do not have nerves, they are rooted in the nail bed, which does have nerves.

Nail and hair follicles are similar in that mitosis takes place in both of these areas. For nails, this mitosis takes place in the nail root. These new cells also produce the protein keratin—similar to new hair cells. However, the keratin is stronger in nails. Once these cells produce keratin, they die. These dead cells make up the nail body; however, the nail bed underneath is a living tissue. The nail bed is under the entire nail.

Thinning and Baldness: Follicles Slow Down as They Age

Hair follicles slow down their growth around the age of 40; therefore, hair production also slows down. This means that hair is not quickly replaced when it naturally falls out, leading to thinning and some patches of baldness.

Baldness, also known as male-pattern baldness, is caused by certain conditions, which include aging. Scientists are not exactly sure what causes baldness, but heredity and androgen (male sex hormone) levels appear to play a role, as well as certain medical conditions. There is evidence that baldness is linked to a gene that begins behaving in a way that changes how the hair follicle responds to circulating hormones.

In addition to the root and the nail bed, there are six other important parts of the nail anatomy:

- *Body*: the main part of the nail, also known as the nail plate

- *Free edge*: the portion of the nail that extends beyond the fingertip

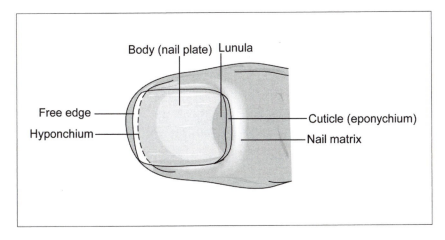

Figure 5.3 Anatomy of a fingernail. (Legger/Dreamstime.com)

- *Eponychium*: Also known as the cuticle, this is a row of dead skin located between the nail plate and the matrix

- *Lunula*: Located right above (and sometimes covered by) the cuticle, this is the crescent shaped, opaque area at the base of the nail plate

- *Hyponchium*: This is the portion of skin located right below the free edge of the nail; this area has an abundant supply of nerve cells, which means it is sensitive to pressure, heat, and other potential sources of pain

Sensory Receptors

This chapter has already included some discussion on the sensory receptors in the integumentary system, particularly in the hair root and nail bed, as well as the skin.

It is important to note that in the skin's dermis layer, there are sensory receptors for the cutaneous senses, including touch, pressure, pain, and temperatures, including cold and heat. Receptors are specific to the senses that they detect, based on their structure. Sensory receptors for pain are called **free nerve endings**, while receptors for the other senses are called **encapsulated nerve endings**. These encapsulated nerve endings are surrounded by a sensory nerve ending.

These receptors communicate with the nervous system about any impact or changes that the environment is having on the skin and its sensory organs. Areas with a high concentration of receptors are the most sensitive. For example, fingertips have a large amount of these receptors in a small area, especially compared to the upper portion of the arm, where the receptors are more spread out. This means that the fingertip will be more sensitive.

Because these receptors detect changes and communicate with the brain, they work closely with the nervous system, which is the focus of Chapter 8.

Glands

In addition to sebaceous glands, which were discussed earlier in this chapter, there are some other important glands in the integumentary system: ceruminous and sweat glands.

SIDEBAR 5.4
Acne

Acne is one of the most common skin disorders, and is related to hormone and other substances plugging the skin's sebaceous glands—also know as oil glands—as well as hair follicles. The plugging of the pores can cause outbreaks of lesions called pimples or "zits." Typically, lesions develop on the face, neck, back, chest, and shoulder.

Medical experts do not routinely describe acne as a disease related to lesions as pimples or zits, but rather as a disease of the pilosebaceous units (PSUs). These PSUs are made up of a sebaceous gland connected to a follicle. While PSUs are found all over the body, they are most populous on the face, upper back, and chest. These glands produce an oily substance known as sebum. This substance typically makes its way to the skin's surface through the follicle's opening, called a pore. Lining the follicle are cells called keratinocytes.

Acne occurs when the sebum, hair, and keratinocytes that occupy the narrow follicle form a plug in the pore, which prevents the sebum from being released to the surface of the skin through the follicle. When the oil from the sebum mixes with the keratinocyte cells, it causes the formation of bacteria known as *Propionibacterium acnes* (*P. acnes*). These bacteria produce various chemicals and also attract white blood cells that cause inflammation. Eventually, the wall of the plugged pore will break down. When this happens, the sebum and bacteria spill on to the surface of the skin, which causes pimples to form.

There are a number of different kinds of pimples or lesions. The basic and most common is called the comedo. This lesion is a plugged hair follicle that is often inflamed. If it stays beneath the skin, a white bump often appears that is called a whitehead. If the comedo reaches the surface of the skin, it is called a blackhead, referring to its appearance on the skin's surface. This black appearance is often thought to be because of dirt, but because of changes in the sebum when it is exposed to air.

Some other acne lesions the following:

* *Papules*: These lesions are inflamed pink bumps that are tender.

* *Pustules*: Also called pimples, these lesions contain a head of white or yellow pus and an inflamed base.

- *Nodules*: These are large and solid lesions rooted deep in the skin and painful to the touch.

- *Cysts*: These lesions are also deep-rotted, and filled with pus. Cysts can often cause scarring.

Ceruminous glands are type of sebaceous gland found in the dermis layer of the ear canal. These glands produce a secretion called **cerumen**, more commonly known as ear wax. While cerumen helps to moisturize the eardrum's outer surface, it can impact the hearing if it builds up and becomes impacted in the ear canal. This impacting can inhibit the eardrum's ability to vibrate and function properly.

There are two types of **sweat glands** in the body: apocrine and eccrine. The apocine glands respond primarily during times of stress and heightened emotion. They are primarily found in the underarm or axillae areas, as well as the genital areas.

The eccrine glands are located all over the body, particularly in the forehead, upper lip, palms, and soles of the feet. These glands secrete through a duct that is in the form of a coiled tube, located in the dermis. This duct connects to a **pore** on the skin's surface.

The eccrine gland and the sweat it produces are important in maintaining the body's temperature. When the body temperature rises, such as during exercise or when in warm temperatures, sweat production is increased, and sweat makes its way to the skin surface. The sweat then evaporates, ridding the body of excess heat. However, this can be dangerous if too much heat is lost. Excess loss of body water through sweat can lead to dehydration, which is a form of heat exhaustion. This is why medical experts recommend that people increase their fluid intake during periods of intense heat and exercise.

Summary

This chapter explored the integumentary system, which is composed of the skin and its accessory organs, which include hair and nail follicles as well as glands and sensory receptors. One of the skin's most important roles is to protect the body from harmful substances, as well as detect changes in the body through sensory receptors. The two layers of the skin are the epidermis (outer) and the dermis (inner). There are two types of glands, the ceruminous glands, which are related to the skin surrounding the ear, and the sweat glands. Sweat glands play an important role in keep the body at a normal temperature.

6

The Lymphatic System

Julie McDowell and Michael Windelspecht

Interesting Facts

- The lymphatic system returns about 3.17 quarts (3 liters) of fluid each day from the tissues to the circulatory system.

- The average macrophage can engulf 100 bacteria a second.

- A plasma cell (B cell) can produce over 2,000 antibodies per second.

- The thymus gland, the site of T cell maturation, reaches its maximum size when a person is age 12, then decreases in size with age.

- A cubic millimeter of blood can contain up to 10,000 leukocytes, of which up to 70 percent are neutrophils and 25 percent lymphocytes.

- In patients with leukemia, the total white blood cell content per 0.03 ounces (1 cubic millimeter) of blood may reach over 500,000.

- Vaccinations against smallpox have been in use since the time of the ancient Chinese civilizations.

Chapter Highlights

- Lymphatic system's cells and chemicals

- Lymphatic fluid

- Lymphocytes

- White blood cells

- Cellular signals and markers in the lymphatic system

- Lymph nodes and circulation

- Lymph vessels

- Bone marrow and thymus

- Spleen, tonsils, adenoids, Peyer's patch, and appendix

- Immune response

- Genetic and acquired immunity

- Vaccines

Words to Watch For

ABO group	Extrinsic factor	Neutrophils
Acquired immunity	Genetic immunity	Normoblast
Afferent vessels	Glycoproteins	Opsonization
Agglutination	Hemolysis	Osmosis
Alleles	Hydrophilic	Phagocytic cells
Anemia	Hydrophobic	Plasma
Appendectomy	Hypoxia	Protease
Appendicitis	Immunity	Reticulocyte
Autoimmune disease	Immunoglobulins	Rh factor
Biomolecules	Intrinsic factor	Tonsils
Bone marrow	Lipoproteins	Toxoid
Chemotaxis	Lymph	Triglycerides
Complement fixation	Lysozyme	Vasodilation
Cytokine	Malignant	Virus
Efferent vessels	Medullary cords	
Erythropoietin	Monocytes	

Introduction

The natural world is an exceptionally hostile place, with pathogenic and parasitic organisms waiting to exploit any weakness in an organism. Fungi, bacteria, parasitic worms, protistans, and viruses abound in the natural world. The concept of survival of the fittest extends from the lowest life forms to the complex environments of primates. In order to survive, all organisms must possess some mechanism of combating invaders. Humans are no exception to this rule. They are in a biological arms race with the microscopic world. Luckily, humankind possesses one of the most elaborate defensive systems on the planet—the lymphatic system.

The primary task of defending the approximately 100 trillion cells of the body against this onslaught of invaders rests with the lymphatic system. While other systems do provide some protection, such as the acids of the stomach and the structure of the skin, it is the job of the lymphatic system to initiate an immune response against invading pathogens. The lymphatic system is the system of the body that is responsible for the immune response. This tiered system of defense utilizes physical barriers, such as the skin, and general defense mechanisms, such as the white blood cells. But perhaps the most significant weapon in its arsenal is the specific defense system. In this aspect of the immune response, specialized cells called lymphocytes detect specific invaders (such as fungi, bacteria, and viruses) and eliminate them from the body. This response can be directed against both free pathogens in the body or against cells that have become infected. As an added protection, the specific response has the ability to "remember" an infection, practically ensuring that you will never be infected by the same organism or virus twice.

The lymphatic system does have other roles in the body. First, it acts as a second circulatory system. The lymphatic system is responsible for returning the fluid from the tissues of the body, called interstitial fluid, to the circulatory system. In this regard, the lymphatic system helps regulate water balance, ensuring not only that the tissues have proper fluids, but that excess fluids do not accumulate in the extremities. The second—often overlooked—role is that of a transport system. The lymphatic system moves fat-soluble nutrients from the digestive system to the circulatory system using a special class of molecules called the lipoproteins.

Unlike other body systems, such as the digestive system and endocrine system, the lymphatic system does not have a large number of organs dedicated to the role of immune response. While there are some, such as the thymus and spleen, the majority of the lymphatic system consists of small ducts, minor glands, and specialized cells located in other body systems, which will be discussed in this chapter.

The Chemicals and Cells of the Lymphatic System

The lymphatic system is a complex group of cells, tissues, and organs that are widely dispersed throughout the human body (Figure 6.1). The lymphatic system has three primary functions. First, its cells are primarily

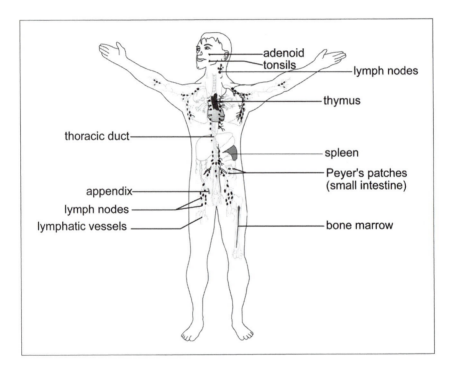

Figure 6.1 Location of the Lymphatic System Organs. In many cases, the organs of the lymphatic system have dual roles in the body. (Sandy Windelspecht/Ricochet Productions)

responsible for the immune response of the body. For this reason, the lymphatic system is frequently called the immune system. Most people are familiar with the immune system as it provides resistance to disease. Modern diseases such as acquired immunodeficiency syndrome (AIDS) and sudden acute respiratory syndrome (SARS) greatly challenge the capabilities of our immune system. Second, the vessels of the lymphatic system actually represent a separate circulatory system in the human body. Unlike the cardiovascular circulatory system, the lymphatic system does not directly supply nutrients or oxygen to the tissues of the body, but rather is primarily involved in the return of fluids from the tissues. Finally, the lymphatic system is involved in the transport of select nutrients from the digestive system to the circulatory system.

These initial sections of this chapter provide an overview of the molecules, cellular components, and chemical signals of the lymphatic system. It focuses primarily on those aspects that are associated with the immune response, although some transport molecules are also discussed. The interaction of these cells, signals, and molecules to create an immune response will be covered later in the chapter.

Subcellular Components of the Lymphatic System

Complement Proteins

As the name suggests, these proteins complement or assist in the function of the immune response. These are nonspecific components of the lymphatic system, meaning that they do not recognize specific types of pathogens entering the body, but instead target any form of invading bacteria or fungus. From an evolutionary perspective, complement proteins probably represent the simplest and oldest form of immune system. Forms of complement proteins are found in all animals. There are approximately 20 different types of complement proteins in humans; collectively, they are called the complement system. Complement proteins move throughout the circulatory system in an inactive form, commonly called a zymogen. The mechanism by which they are activated is dependent upon the class of the complement protein. Although complement proteins are found in the circulatory system, they are considered to be part of the lymphatic system due to their association with the immune response.

The complement proteins may either target invading fungal and bacterial cells directly, or they may be recruited by antibodies or other cells of the immune system. The proteins have a variety of functions. Some classes are involved with attacking the membrane of the invading pathogen causing it to lyse, or break. Other classes interact with the antibodies secreted by the B lymphocytes (see the "Lymphocytes" section later in this chapter). This is often called the classical pathway, because it is the most common mechanism of complement system activation. Once activated by an antibody, the complement proteins form a pore in the membrane of the invading cell, causing it to lyse.

Some complement proteins act as molecular flags. This class sticks to the surface of the pathogen, but rather than causing the membrane to rupture, these proteins signal macrophages and other **phagocytic cells** of the immune system to envelop the invading cell and destroy it. Other classes of complement proteins are involved in the inflammatory response or in activating enzymes in the blood.

The complement proteins that directly lyse the membrane of the pathogen do so by what is called the alternative pathway. In this case, the inactive proteins are activated by some component of the bacterial or fungal cell wall. Once activated, the proteins congregate on the invading cell and form a pore through the membrane, disrupting the membrane barrier of the cell and causing it to lyse.

While complement proteins may appear to be an effective mechanism of immune response, they lack the ability to target specific types of cells that are invading the body. The task of targeting specific invaders falls to the cells of the immune system.

Chylomicrons and Lipoproteins

One aspect of the lymphatic system that is not involved in the immune response is the transport of fat-soluble material from the digestive tract. This includes not only **triglycerides**, but also the fat-soluble vitamins. These **hydrophobic** molecules are packaged within the small intestine into spherical structures called **lipoproteins**.

Lipoproteins are a combination of fats and proteins. Following enzymatic digestion in the lumen (cavity) of the small intestine, fatty acids are reassembled into triglycerides in the epithelial cells of the small intestine. They are then packaged into chylomicrons. Chylomicrons represent one form of lipoprotein that is manufactured within the lining of the small intestine. Due to their size and hydrophobic characteristics, chylomicrons cannot pass into the capillaries within the villi of the small intestine, and thus are unable to be transported to the liver in the same manner as the majority of nutrients. Instead, they enter into the lacteals of the digestive tract.

Once in the lacteals of the intestines, the chylomicrons utilize the lymphatic system to bypass the liver and travel to the heart via the thoracic duct, where they enter into the bloodstream. At this point the vitamins and energy-rich nutrients within the chylomicron are removed by the tissues, and the chylomicron becomes an empty shell. The other lipoproteins, such as low-density lipoproteins (LDLs) and high-density lipoproteins (HDLs), are manufactured by liver tissue and do not enter the lymphatic system.

Antimicrobial Proteins

The surface cells of the body, called the epithelia, are most often the first to experience an attack by an invading organism. For this reason, many of the body's surfaces secrete antimicrobial proteins or enzymes. An enzyme is a chemical compound (usually a protein) that accelerates a chemical reaction. Although enzymes are most often thought of in association with the digestive or nervous systems, in fact they are active in all of the systems of the body.

The surfaces of the eyes and mouth, because they are moist environments and warmer areas of the body, represent an ideal location for a microbial attack. At these locations, the body secretes an enzyme called **lysozyme** in the saliva and tears. Lysozyme acts by degrading the cell walls of invading bacteria. Because animal cells lack cell walls, they are not disturbed by the presence of the enzyme.

This is not the only example of antimicrobial compounds in the body. Technically, the **protease** enzymes of the stomach may be considered a part of the immune response, because, in cooperation with the hydrochloric acid of the stomach, they inhibit the activity of pathogenic organisms. In the small intestine, specialized cells called Paneth cells secrete an antimicrobial compound called cryptidin. Even the bacteria located within the large intestine assist with patrolling against incoming pathogens. *Escherichia coli* (commonly called just *E. coli*), frequently considered to be a pathogen itself, helps protect the large intestine by secreting a chemical called colicin that prevents growth of pathogenic organisms.

These antimicrobial systems are not designed to completely prevent an attack by a pathogenic organism. Instead, like the complement proteins, the antimicrobial substances noted in this section act to slow the growth of an invader and give the specific defense mechanisms (lymphocytes) time to prepare. In this regard, antimicrobial systems are very effective in their mode of action.

Lymphatic Fluid

The fluid content of the lymphatic system is actually derived from the circulatory system. In the circulatory system, the capillaries represent the location where gas and nutrient exchange is most likely to occur with the surrounding tissue. Capillaries are fragile structures, whose walls are typically only one cell thick. However, these cells, called endothelial cells, do not form a solid structure, like that of a hose. Instead, there are small pores between the cells that form the lining of the capillaries. These pores are too small to allow the cells and plasma proteins of the circulatory system to pass, but large enough to allow a free exchange of fluid with the surrounding tissues. This fluid represents the medium through which nutrients and gases may be exchanged. The fluid, called interstitial fluid, bathes most tissues of the body. Cells typically deposit waste in the interstitial fluid for pickup by the circulatory system, and receive nutrients and gases to conduct their metabolic processes.

The majority of this fluid is reabsorbed back into the capillaries. However, this process is only about 85 percent effective. Each day, about 3.17 quarts (3 liters) of fluid is not reabsorbed back into the capillaries, but instead remains in the tissue. This amount may not sound significant, but in an average adult, there is only 5.28 quarts (5 liters) of blood. It would seem that the loss of fluid from the capillaries would represent a severe challenge for the circulatory system, and the organism as a whole.

The lymphatic system makes up the difference by recycling the interstitial fluid and returning it back to the circulatory system. In most people, the lymphatic system returns around 3.17 quarts (3 liters) of fluid daily. In other words, the output of the circulatory system to the tissues is matched by the input of interstitial fluid from the lymphatic system.

Lymphatic fluid does not contain red blood cells, and in general lacks any pigmentation. However, despite its lack of color, there are plenty of ions, molecules, and cells in lymphatic fluid. These include ions such as sodium (Na^+) and potassium (K^+), chylomicrons, and a host of cells associated with the immune response (see next section).

Cells of the Lymphatic System and Immune Response

The immune system utilizes a number of different cell types to protect the body from infection. The major classes of cells are listed in Table 6.1.

TABLE 6.1
Cells of the Immune System

Cell class	Type	Specific or nonspecific defense	General function
Lymphocytes	Natural killer (NK)	Nonspecific	Targets virus-infected cells
	T cells	Specific	Attacks antigen-presenting cells
	B cells	Specific	Produces antibodies to attack free antigens
White blood cells	Macrophages	Nonspecific	General phagocytic cells
Neutrophils		Nonspecific	One-time-use cells that contain powerful chemical reactions
Eosinophils		Nonspecific	Destroys parasitic organisms
Basophils		Nonspecific	Releases histamine
Mast cells		Nonspecific	Releases histamine

These cell types may either be generalists (nonspecific defense mechanisms) or specialize in the destruction of certain identified invaders of the body. Cells of this second class belong to the specific defense mechanism of the body. The lymphatic cells are derived from the same type of cell in the bone marrow as the cells of the circulatory system. The common name for this type of cell is called a stem cell. While stem cells are commonly thought of as being able to form any type of cell in the body, in reality they vary in this ability. Some stem cells, such as those in early embryonic development, are totipotent, meaning that they have the ability to form virtually any cell type. However, shortly after the embryo starts to develop, stem cells lose their potency, and thus their ability to form certain types of cells. The stem cells in the bone marrow are pluripotent cells, indicating that they are limited in what types of cells they can differentiate into.

The type of stem cell that gives rise to the immune cells, as well as the majority of cells in the circulatory system, is called a hematopoietic stem cell. From this stem cell are derived progenitor cells, which possess an additional layer of specialization. Two different types of progenitor cells are involved with the formation of lymphatic cells. The lymphocytes are derived from the lymphoid progenitor cell, while the leukocytes and macrophages are derived from the myeloid progenitor cell. The cell types also differ in where they mature in the body and their contributions to the function of the lymphatic system and immune response.

Natural Killer Cells

Natural killer (NK) cells are another example of a nonspecific defense mechanism in the body. While the complement system acts nonspecifically against invading fungal and bacterial cells, the role of the NK cells is to eliminate cells of the body that have either been invaded by **viruses** or are cancer cells. It is important to note the difference between NK cells and a form of T cells called the cytotoxic T cells (see the next section). While both attack viral infected cells and cancer cells, cytotoxic T cells are specific in their targets, meaning that they will destroy only cells that have been infected with a specific virus. NK cells are generalists and will destroy any viral infected cell that they come in contact with. Natural killer cells are not phagocytic cells, but rather destroy the target cell by lysing the membrane.

Natural killer cells belong to a class of cells called the lymphocytes, of which the T and B cells are the most commonly recognized. NK cells are formed in the same manner as T and B cells (see the next section), but do not mature in the thymus, as is the case with the T cell.

Lymphocytes
The term lymphocyte is most commonly used to describe two groups of lymphatic cells, the B cells and T cells, although, as noted in the preceding section, the NK cells also belong to this class. Lymphocytes start as hematopoietic stem cells in the bone marrow of the long bones of the body. The hematopoietic stem cells form progenitor lymphoid cells, which then divide into the cell lines that will form the B cells, T cells, and NK cells. Unlike B and T cells, NK cells do not require additional processing and instead proceed directly into action as nonspecific defense mechanisms.

B and T lymphocytes are named for the location in the body in which they complete their maturation process. An immature T cell migrates to the thymus to finish its development; a B cell completes its maturation in the bone marrow. The "B" does not actually stand for "bone," but rather a structure called the bursa of Fabricus. This structure is only found in birds, but is where the B cells were first discovered. Since the B cells of all other vertebrates mature in the bone marrow, the B is commonly considered to refer to "bone." Although both B cells and T cells are lymphocytes and are involved in the defense of the body against specific pathogens, their modes of action are very different.

T and B cells both respond to specific antigens in the human body. An antigen is a molecule that invokes an immune response. All cells and viruses have unique antigens present on their surface. What distinguishes the cells of our body from invading viruses and the cells of invading bacteria, fungi, protistans, and parasitic worms is the presence of self-markers. In other words, if a cell cannot identify itself as a normal part of the human body, it runs the risk of initiating an immune response.

The role of the B cells is to develop antibodies against antigens that present themselves in the tissues and fluids of the body. Antibodies are proteins that target the antigen and either mark it for destruction by nonspecific mechanisms or physically destroy the molecule. (The action of antibodies will be covered in greater detail later in this chapter). Because almost anything may be an antigen (proteins, cellular debris, chemicals, etc.), it is possible for a B cell to mistakenly identify a cell of the body as an antigen. Therefore, immature B cells are screened while in the bone marrow before maturing and being sent to secondary lymphoid tissues, such as the appendix and lymph nodes. This process is often called self-tolerance, because the B cell must be able to tolerate the wide range of potentially false antigens that are produced by the cells of the body. However, sometimes this screening is not completely effective, resulting in an **autoimmune disease**.

Each B cell of the body will recognize—and produce antibodies against— one specific antigen. Antibodies are proteins, and are manufactured in the same manner as other proteins in the body. The instructions for producing the antibody are stored as genes in the deoxyribonucleic acid (DNA). When needed, these genetic instructions are transcribed into a message (called messenger ribonucleic acid, or mRNA), which proceeds to the cytoplasm of the cell to be translated into a functional protein.

All antibodies have a characteristic structure (Figure 6.2). Each antibody contains two light chains and two heavy chains. They are called the light and heavy chains based upon the number of amino acids in the peptides that make up their structure. The chains are held together by disulfide bonds, forming a "Y"-shaped molecule. There is very little variation in the constant regions (or C regions) of the heavy chains in the antibody.

In humans, there are just five major variations in this area of the heavy chain, which correspond to the five major classes of antibodies. This combination of light and heavy chains is responsible for the tremendous variation in antibody specificity. At the terminal end of each chain is an area called the hypervariable segment, which is ultimately responsible for targeting a specific antigen. The mechanism by which these hypervariable regions in the chains are generated is one of the amazing features of the immune system.

The structure of the antibody, and thus its effectiveness in recognizing the correct antigen, is determined by the sequence of genetic information that is used to construct the protein. Even though the human genome

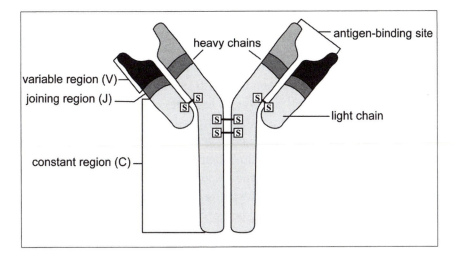

Figure 6.2 Antibody Structure. This diagram illustrates the basic structure of an antibody protein. (Sandy Windelspecht/Ricochet Productions)

contains over three billion pieces of information, organized into over 30,000 genes, there is still not enough information or room to have a single gene for each antibody that a human being may require over an entire lifetime. It is believed that the immune system has the potential to produce between one million and one billion different types of antibodies. It is simply not possible that each antibody is the result of a single dedicated gene in the DNA.

It is now known that there are only a few hundred genes that are responsible for generating the diverse array of antibodies. These genes are grouped into four major types:

- Genes grouped into the C regions are responsible for generating the protein sequences in the constant regions of the heavy and light chains.

- The J genes are responsible for generating the small peptide segments that link portions of the antibody together.

- The D group of genes encodes for a small diversity region found only in the heavy chain.

- The V genes provide the information for generating the hypervariable segments of the light chains.

It is important to note that the following process of selecting the genes that will form an antibody occurs before the B cell encounters an antigen. In other words, as the B cell matures, it becomes specific in what type of antigen it will respond to. It is extremely possible that a B cell may never come in contact with its antigen, and thus never be involved in an immune response. However, the large number of B cells in circulation, and the mechanism of the specific immune response, means that the body needs only one B cell that recognizes the antigen to mount an effective immune response. Because many invading pathogens, such as bacteria and viruses, may present multiple antigens, it is possible that more than one B cell may be producing antibodies for the same pathogen at the same time.

To construct a light chain, it is necessary to have a single V, J, and C gene segment (there are no D segments in the light chain). There are 30 to 40 V segments, four to five J segments, and a single C segment in the area of DNA responsible (located on human chromosomes 2 and 22) for the formation of the light chain. This alone produces around 200 different combinations. Furthermore, the regions are subject not only to mutation, but also to minor variations in the reading and processing of the genetic information. These mistakes serve to increase the variation in the segments. Thus, it is highly unlikely that two B cells will be identical in the types of antibodies that they produce. Before maturation, the lymphocyte possesses all of the gene segments. By a process called somatic recombination, a single V and J region are selected and then matched up to the one C region. After translation, the result is a protein (also called a peptide) that will become the light chain of the antibody.

The synthesis of a heavy chain is slightly different, but follows the same general pattern. There are 65 V segments, 27 D segments, and 5 J segments available on chromosome 14 to construct a single heavy chain, resulting in 10,530 possible combinations. Once again, a single V, D, and J segment are combined, and then linked to a C segment. The same errors that produced variation in the light chains may also play a role in the formation of the heavy chains, resulting once again in an almost endless source of variation in heavy chain structure. Following processing, transcription of the information, and translation into a functional protein, the result is the heavy chain of the antibody. The light chains and heavy chains are then linked together, forming an antibody.

Although all B cells undergo a similar maturation process, there are minor variations in their form and function in the immune system. Once activated by a specific antigen, a B cell rapidly divides, forming a large number of effector B cells, or plasma cells. These cells actively combat the antigen in the body in what is called the primary immune response. As the primary response progresses, some of the activated B cells are retired, forming memory B cells. These cells are responsible for the secondary immune response that occurs when the body is exposed to the same antigen later in time. All B cells are involved in the humoral response, which targets free antigens in the system.

As noted, T cells complete their maturation in the thymus, one of the primary lymphoid organs of the body. The thymus is located just above the heart. In the thymus, the immature T cells undergo a series of modifications. The most important of these changes occurs as specific genes within the T cells are activated, which enables the production of unique proteins, called **glycoproteins**, on the surface of these cells. These proteins, examples of which are called CD4 and CD8, play an important role in the function of the immune system. As was the case with the B cells, the maturing T cells in the thymus are screened to ensure that they are not recognizing any of the tissues or cells of the body as invading antigens. Those T cells that display an affinity for self are targeted for cell death and usually do not mature.

While T cells are specialized for the targeting of antigen-presenting cells of the body, there are actually several different forms of T cells, each of which has a specific function in the immune response:

- *Helper T cells.* Helper T cells serve as the liaison between the nonspecific and specific defense mechanisms. They are responsible for activating both the humoral and cell-mediated responses in the body, by interacting with both mature T and B lymphocytes.

- *Cytotoxic T cells.* These cells destroy infected body cells under the direction of the helper T cells.

- *Inducer T cells.* Cells located in the thymus that are responsible for T cell maturation.

- *Suppressor T cells.* The task of these cells is to shut down the immune response once the infection is complete. These cells divide much slower than other T cells, producing a natural delay mechanism.

- *Memory T cells.* As was the case with the B cells, some T cells are held in reserve for future exposures to the antigen. These are called memory cells, because they originated with the initial exposure to the antigen.

The activity of T cells is highly dependent on cell-to-cell signaling using proteins embedded in the plasma membrane. T cells not only serve to identify cells of the body, but also to distinguish between infected and healthy cells so as to prevent unnecessary tissue damage. This identification is made possible by specific receptors on the surface of the T cell. The antigen identification portion of these receptors is highly variable. T cells use a similar method of generating variation, as do the B cells with antibody formation. Within the genome of the T cell are the same variable (V), diversity (D), joining (J), and constant (C) families of genes that are present in the B cells. These genes are rearranged in much the same way as in the B cells to produce a receptor that is specific for one type of antigen recognition. As is the case with the B cell, each T cell is specific in what it can recognize. The difference is that while the antibodies of the B cell recognize free antigens, such as what might be found in the fluids of the body, the receptors of the T cells are designed to identify cells that are presenting a specific antigen as the result of being infected by a pathogen, such as a virus. This is a complex interaction, which involves a number of cellular markers and proteins.

White Blood Cells
The term white blood cell is used as a general description for a wide variety of cells in the immune system, including macrophages, neutrophils, eosinophils, and basophils (the general function of each is provided in Table 6.1). A white blood cell begins as a hematopoietic stem cell in the bone marrow. This cell then begins to differentiate into a myeloid

progenitor cell, which then becomes basophils, eosinophils, **neutrophils**, and a precursor of macrophages called **monocytes**. Typically, the term white blood cell indicates the macrophages.

Macrophages are the general workhorses of the immune system. These are amoeba-like cells that move throughout the body. Unlike many cells of the circulatory and lymphatic system, the macrophages are not confined to capillaries. Instead, they are able to move freely between the circulatory system and the interstitial fluids that bathe the tissues of the body. Macrophage is the collective name for these cells, but they are sometimes called by other names throughout the body (see Table 6.2). Inactive macrophages are monocytes. These may be activated at the site of an infection by a variety of mechanisms.

Macrophages are phagocytic cells, meaning that in order to destroy pathogens, macrophages must first ingest them. Phagocytosis involves a budding in of the cell membrane, forming a vesicle inside the cell. Once the pathogen is engulfed, the macrophage can utilize two major mechanisms for destroying the invader. First, the macrophage may merge the pathogen-containing vesicle with another cellular vesicle called a lysosome.

The lysosome contains powerful digestive enzymes that effectively "eat" the pathogen, rendering it harmless. A second mechanism involves making the internal environment of the vesicle containing the pathogen extremely toxic. Macrophages can produce free radicals such as nitric oxide and superoxide anion. These chemicals are highly destructive to organic molecules, and quickly destroy the incoming pathogen.

TABLE 6.2
Nomenclature for Macrophages in the Body

Tissue	Name
Digestive system (liver)	Kupffer's cells
Urinary system (kidney)	Mesangial cells
Connective tissue	Histocytes
Respiratory system (lungs)	Alveolar macrophages
Nervous system (brain)	Microglial cells

Once the pathogen has been destroyed, an interesting change happens in the receptors of the macrophage. The macrophages are nonspecific generalists, but they play an important role in informing the lymphocytes (specific defense) of the presence of a pathogen in the body. Small pieces of the pathogen are moved to receptors on the cell membrane. These receptors, part of the self-identification process (see the next section, "Identification of Self: The Role of Cellular Markers"), act as an activation mechanism for the helper T cells, an important link between the specific and nonspecific responses. Once the macrophage alters the surface receptors, it is then called an antigen-presenting cell (APC).

In many ways, the neutrophils are very similar to the macrophages. Both are phagocytic cells that patrol the body looking for pathogenic organisms or viruses. Neutrophils are actually the most abundant type of white blood cell in the body. They are typically found only at the site of an infection, usually because they are attracted by macrophage activity and chemicals released from damaged cells. Unlike the macrophages, which rely almost exclusively on phagocytic activity, the neutrophils have a variety of options available for the destruction of incoming pathogens.

Neutrophils may engulf pathogens in a manner similar to the macrophages. However, they also possess a more destructive mechanism. Each neutrophil contains a limited number of internal structures, called granules. These granules contain a variety of substances that are highly toxic to microorganisms. These include oxygen radicals, antimicrobial proteins, peroxynitrate (a nitric oxide compound), and hydrogen peroxide. When a neutrophil reaches the site of an injury or infection, it releases these chemicals into the surrounding environment, killing not only microorganisms, but often cells of the body as well. Each neutrophil only contains a limited arsenal of chemicals, and once depleted, the neutrophil dies. Often, the neutrophil may be destroyed by the very chemicals that it releases. The debris from these dead cells is what forms the pus at a site of a wound or infection.

Another form of white blood cell is the eosinophil. The eosinophils are another nonspecific mechanism, but one that is targeted against parasitic organisms, such as intestinal worms, flukes, microscopic nematodes, and even ticks and mites. Eosinophils act by orientating themselves along the surface of the parasitic invader and then releasing chemicals to destroy the membrane or surface of the organism.

The last two major types of white blood cells are not directly involved in the destruction of invading organisms, but rather are involved in some of the chemical signaling. These are the basophils and mast cells, which are responsible for the release of a chemical called histamine, an important chemical in the inflammatory response. The role of these cells in histamine production is discussed later in this chapter.

Identification of Self: The Role of Cellular Markers

Before beginning a discussion of cellular markers, it is first important to understand the basic structure of a cell membrane. The membrane of a cell is composed primarily of a molecule called the phospholipid. Phospholipids belong to a class of **biomolecules** called the lipids and are unique in that they contain both **hydrophilic** and hydrophobic regions. When these molecules are placed in an aqueous environment, the hydrophilic and hydrophobic regions align to form a double layer, called a lipid bilayer. This lipid bilayer forms the basic structure of the cell membrane, the properties of which effectively block the passage of most molecules into the cell.

Located within the phospholipid layer of the cell membrane are a wide variety of proteins. Some of these proteins serve as channels through the membrane, while others act as receptors for chemical signals passing back and forth between the tissues of the body. The proteins of interest in the immune response belong to the glycoproteins, the proteins that have a sugar group attached to their outer surface. Glycoproteins are common on the surface of the cell membrane, but two types of glycoproteins play an important role in the immune response. These are the major histocompatibility complexes (MHC) markers.

MHC markers should be considered a form of cellular identification card. All of the cells of the body have the same identification tag, which enables the body to distinguish between "self" and invading microorganisms and viruses. There are two classes of MHC markers in the cell, called MHC-I and MHC-II. MHC-I markers are more general, and are found on every cell of the body. MHC-II markers are slightly more specific, and are found almost exclusively on cells of the immune system.

There is a tremendous amount of variation between individuals in the structure of the MHC markers. While only three genes are responsible for forming an MHC-I protein (all on human chromosome 6), there exists a number of variations, or **alleles**, for each of these genes. This allele effectively ensures that the MHC signature of one individual is unique.

Chemical Signals in the Lymphatic System

Because the cells of the immune system are not all localized in a single tissue, the system possesses an elaborate series of chemical signals to operate effectively. These signals perform a variety of functions, from the recruitment of nonspecific defense mechanisms to the activation of cells involved in the identification and destruction of specific pathogens. The cytokines are a special group of chemical signals in the immune system. These protein, or peptide, signals are secreted primarily by the T cells of the body to influence the activity of both the specific and nonspecific cells of the immune system. There are currently 13 different types of cytokines that have been identified. The majority of these belong to two major groups, the interferons and the interleukins (Figure 6.3).

As their name implies, the interferons are involved in an interference response in the immune system. Several different types of interferons are active in the immune response. Type 1 interferons (also called interferon alpha and interferon beta) are used as a local defense against invading viruses. When a cell is infected with a virus, it secretes interferons into the surrounding interstitial fluids. On neighboring cells, the interferon interacts with a receptor common on the surface of all cells. This causes the cell to activate antiviral protection mechanisms (frequently protease enzymes), which inhibit viral replication in uninfected cells. The type 1 interferons also enhance the development of APCs, and serve to activate NK cells in the area. It is important to note that the secretion of interferons does not protect the infected cell, only those in close proximity to it. It is also nonspecific, meaning that neighboring cells are temporarily protected against any viruses in the area.

Another type of interferon is called interferon gamma. This interferon is secreted by selected T cells and is not directly related to the type 1 interferons. The role of this interferon is to activate macrophages near the cytotoxic T cells, thus providing a better coverage in the area of an infection or wound.

Interleukins are the second major form of cytokine. There are a number of interleukins in the immune system, the most common of which are

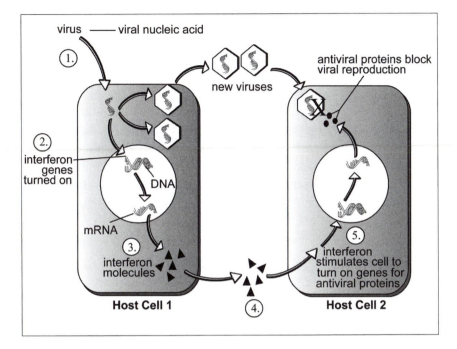

Figure 6.3   Interferon Activity. The cell on the left has been infected by a virus. It then produces interferon, which is detected by a nearby cell (on the right), enabling the production of antiviral proteins. (Sandy Windelspecht/Ricochet Productions)

interleukin-1 (IL-1) and interleukin-2 (IL-2). IL-1 acts as a link between the nonspecific and specific defense systems. After a macrophage has engulfed a pathogen and become an APC, it secretes IL-1 to help activate helper T cells in the area. The helper T cells then communicate directly with cytotoxic T cells and B cells to begin the humoral and cell-mediated responses to the antigen.

The second major form of interleukin, IL-2, is basically an activated signal. Secreted by helper T cells, the signal activates B cells to begin antibody production, as well as cytotoxic T cells to begin destruction of infected cells of the body. The loss of helper T cells, as is the case with AIDS, means that this signal is not present, and the specific defense systems are not activated.

There are other chemical signals in the immune system besides the interferons and interleukins. In response to injury, basophils and mast cells release a chemical called histamine. Histamine causes the cells of the capillary beds (circulatory system) to dilate, increasing the amount of fluid (but not the amount of red blood cells) flowing out of them. This increase in interstitial fluid increases the pressure, slowing the spread of bacteria and other pathogens into the wound. In addition, clotting proteins can now move more easily to the site of the wound, allowing for a more rapid healing process. Macrophages and NK cells also benefit by the ease of moving into the interstitial spaces, allowing for a more rapid cleanup to begin. Antihistamine medications, such as those used for the common cold, reverse this process and may actually slow the immune response.

An Examination of the Lymph Nodes and Lymphatic Circulation in the Body

One of the primary functions of the lymphatic system is to capture and collect the protein-rich fluid that escapes from the circulatory system's blood vessels and deposit it back into the tissue network. These proteins cannot be reabsorbed; therefore, the lymphatic circulatory system must fetch them and bring them back. As explained previously, the lymphatic system's circulatory functions are separate from the cardiovascular system in the human body. However, unlike the cardiovascular system, the lymphatic system's circulatory processes are focused on returning fluids from other areas of the body, rather than directly supplying the tissues with the nutrients and oxygen that they need to function.

This section will begin to look at some of the basic components of the lymphatic system—the lymph nodes and the elements involved in lymphatic circulation, in addition to the system's primary organs, the **bone marrow** and the thymus. These aspects of the lymphatic system are important to understand before learning how the more complex functions, such as immunity and autoimmune response, work to protect the body.

Lymphatic Vessels

The cardiovascular, circulatory system and the principal parts of the lymphatic system interact primarily through the lymph nodes, lymphatics,

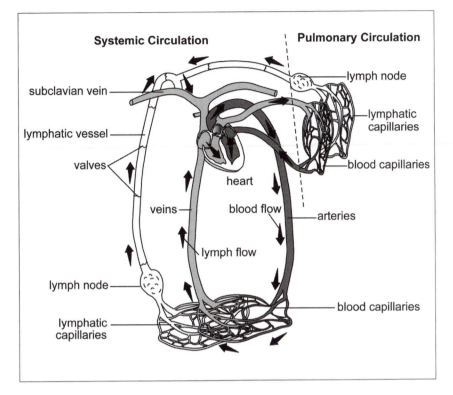

Figure 6.4   The Interaction between the Body's Lymphatic and Cardiovascular Systems. Tissue fluid is collected by lymph capillaries and is then returned to the blood. The arrows indicate the flow of the lymph and blood. (Sandy Windelspecht/Ricochet Productions)

and lymph capillaries (Figure 6.4). The lymphatic vessels begin as blind-end tubes—called lymph capillaries—that form in the spaces between cells.

Lymph capillaries are slightly larger, in addition to being more permeable, than the blood capillaries in the circulatory system. These capillaries can form in most regions of the body, and converge to form larger lymph vessels called lymphatics. These lymphatic vessels have a vein-like appearance, although their walls are thinner and they contain more valves than blood veins. In addition, at various spots in their structure, lymphatics contain lymph nodes.

Lymph is the name of the fluid that enters the lymph capillaries. As explained previously, tissue fluid comes from the filtration in the capillaries. While the process of **osmosis** allows much of this fluid to return to the blood, some of the fluid is lodged in interstitial spaces. The lymphatic vessels return this interstitial fluid to the blood to become **plasma** again. Without this occurring, blood volume and blood pressure would rapidly decrease, eventually leading to serious health threats, such as a heart attack or stroke.

Unlike the circulatory system, there is no pump for the lymph. In the circulatory system, the heart serves as the pump to keep blood moving throughout the body. In the lymphatic system, the lymph is kept mobile through the muscles of the lymph vessels (Figure 6.5). As the smooth muscle layer of the larger lymph vessel constricts, the one-way valvular structure prevents the backflow of the lymph.

As the lymph capillaries form lymphatics, the lymphatics eventually merge into two main structures or channels, called the thoracic duct (also called the left lymphatic duct) and the right lymphatic duct. Lymphatic vessels from the left side of the head, neck, and chest, in addition to the left upper extremity and the entire body below the ribs, all converge in front of the lumbar section of the vertebrae to form the cistern chyli vessel, which then continues to creep up the backbone as the thoracic duct. This duct then empties the fluid into the left subclavian vein, where a pair of valves is located to prevent the passage of blood into the thoracic duct. The second channel, the right lymphatic duct, takes the lymph from the right side of the body and then deposits it into the opposite, or left subclavian vein.

When doctors need to take a detailed look at the lymphatic vessels and organs, they rely on a procedure known as a lymphangiography. The lymphatic vessels and organs are filled with an opaque substance and then filmed, which produces a lymphangiogram. This image is useful for identifying edemas, carcinomas, and viewing any irregularities of the lymph nodes.

Lymph Nodes

The lymphatics contain structures that are oval in shape called lymph nodes. These bean-like organs can range in size from 0.04 to 1 inch (1 to 25 millimeters). Blood flows into lymph nodes on the way to subclavian

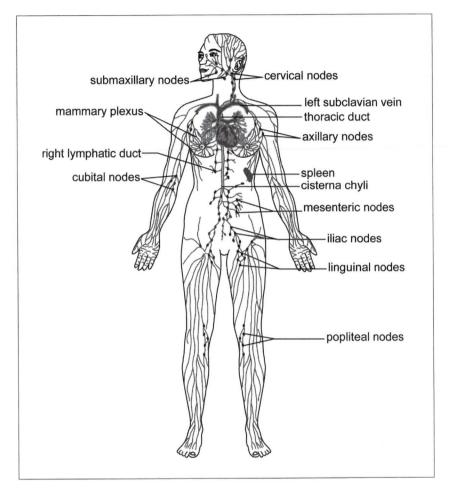

submaxillary nodes

cervical nodes

mammary plexus

left subclavian vein

thoracic duct

axillary nodes

right lymphatic duct

cubital nodes

spleen

cisterna chyli

mesenteric nodes

iliac nodes

linguinal nodes

popliteal nodes

Figure 6.5   Primary Groups of Lymph Nodes and the Lymph Vessel System. The right and left subclavian veins return lymph to the body's blood supply. (Sandy Windelspecht/Ricochet Productions)

veins. Each lymph node contains a hilum, which is a slight depression on one side where the blood vessels enter and leave the node (Figure 6.6).

Three structural elements form the framework of a lymph node: the capsule, the trabeculae, and the hilum. The capsule is made up of fibrous connective tissue that not only covers the node, but also extends into it.

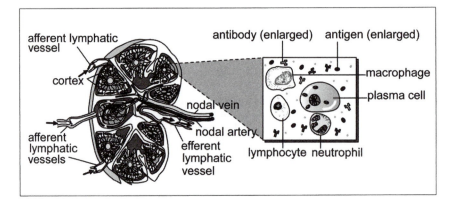

Figure 6.6 The Lymph Node. On the left, a cross-section view of a lymph node, with the arrows indicating the lymph flow. On the right, a detailed view of antigen being destroyed within the lymph node. (Sandy Windelspecht/Ricochet Productions)

These extensions into the node are call trabeculae. Inside the node, the outer cortex is composed of tightly packed lymphocytes organized into lymph nodules. These nodules contain germinal centers, where lymphocytes are actually produced. Then the inner portion of the lymph node is called the medulla, which contains lymphocytes that are organized into strands called **medullary cords**.

Lymph nodes contain two kinds of vessels: afferent and efferent. Lymph leaves the node through one or two **efferent vessels**, while it enters through one or a couple of **afferent vessels**. Once the lymph is in the node, any bacteria and other foreign materials it is carrying are phagocytized (consumed) by macrophages. The lymphocytes contain plasma cells that produce antibodies to counteract any pathogens that the lymph brings with it. These antibodies, in addition to the lymphocytes, will eventually travel to the blood. (Antibodies will be discussed later in this chapter.)

Once lymph enters the node through the afferent vessels, which are located at various places on the surface of the node, the fluid then enters the node's sinuses, which are a series of irregular channels. After passing through the afferent vessels, lymph enters the cortical sinuses and then circulates through the medullary sinuses located between the medullary cords. After the lymph passes through these sinuses, it then travels to the efferent vessels, which are located at the node's hilum structure. While the afferent vessels open only toward the node, the efferent vessels open only outward away from the node, pushing lymph out from the structure. In addition, there are fewer efferent vessels (although the actual vessels are wider) than there are afferent vessels.

Macrophages that contain phagocytic cells are located along the sinuses. Lymph travels through the nodes, and is then processed by these phagocytic cells, which work to separate out the bacteria, dirt, and other contaminants from the fluid. In addition, the nodes are where the lymphocytes and plasma cells are produced, which then lead to the formation of antibodies. When too many pathogens and microbes enter the node, it can become infected, causing the node to enlarge and become inflamed.

Numerous groups, or chains, of lymph nodes are located along the body's lymph vessel network, but the three primary groups of nodes classified according to their location in the body—cervical, axillary, and inguinal—are detailed in Table 6.3. Each of these node groups are located at an important junction of the body: the cervical nodes are located near the neck and head junction; the axillary nodes are located near where the arm meets the trunk of the body; and the inguinal nodes are found near where the leg meets the trunk. This is important because the skin is more likely to break in these areas, thus allowing pathogens to enter the body. For example, it is more likely that skin will break in the head, arms, or legs, rather than the

TABLE 6.3
Groupings or Chains of Nodes

Group	Location and function
Deep cervical lymph nodes	Located along the internal jugular veins, these nodes process lymph from the head and neck.
Axillary lymph nodes	Located in the chest and underarm areas of the body, these nodes process lymph from the skin and chest muscles, which include the breasts.
Inguinal lymph nodes	Located in the groin region, these nodes drain lymph from the lower extremities of the body, including the genitals.

trunk of the body. Therefore, if pathogens enter the body through any of these locations, the lymph will destroy them before they reach the trunk, and also before the lymph is returned to the blood contained in the subclavian veins.

As these lymph nodes are processing the pathogens and bacteria, inflammation and temporary infection can occur. For instance, swollen glands frequently accompany "strep throat," which is an infection caused by a bacteria called Streptococcus. The glands that are swollen are actually the cervical nodes, which have temporarily enlarged as the macrophages fight off the bacteria in the throat's lymph.

There is also a specific kind of lymphatic tissue found in all mucous membranes, which line those systems of the body that have exterior openings to the environment. These include the respiratory, digestive, urinary, and reproductive tracts. All of these systems are lined with mucous membranes for protection. Located under the epithelial layer of these membranes are small groupings of lymphatic tissue know as lymph nodules.

This is an important area for lymph nodules because, while these systems are shielded from some contamination through the mucous membranes, they are still vulnerable to attack by microbes, bacteria, and various pathogens. If bacteria is inhaled and enters the body through the respiratory system, the lymph nodules located in the trachea will counteract that bacteria before it even reaches the blood. Two kinds of lymph nodules are Peyer's patches (located in the small intestine) and **tonsils** (located in the pharynx).

Bone Marrow

The three kinds of blood cells—white blood cells (WBCs), red blood cells, (RBCs), and platelets—are produced in two kinds of hemopoietic tissues: red bone marrow (or simply bone marrow), and lymphatic tissue that is found in the spleen, thymus gland (see the "Thymus" section later in this chapter), and lymph nodes. The red bone marrow is spongy tissue found in flat and irregular bones. Basically, the purpose of the RBCs is to carry oxygen throughout the body. Through a protein that they carry called hemoglobin, RBCs are able to bond to oxygen molecules. (The function of the RBCs are covered extensively in Chapter 2 on the circulatory system, and WBCs are extensively covered earlier in this chapter).

Before RBCs are produced in the bone marrow, they are stem cells that are constantly changing to form all kinds of blood cells. The rate of RBC production is high; approximately a few million are produced every second. This production rate, however, is regulated by the presence of oxygen. If plenty of oxygen is available in the body, then the bone marrow will produce the RBCs at a normal rate. However, if the body is low on oxygen (or in a state of **hypoxia**), then the kidneys will begin producing a hormone called **erythropoietin**, which causes the bone marrow to produce more RBCs, which are then able to carry oxygen through the body. Once the oxygen begins making its way through the circulatory system, the body is no longer in a hypoxic state. A person can become hypoxic following a hemorrhage or other injury that has caused them to lose a lot of blood, or if they spend a significant amount of time in higher altitudes.

As stem cells develop into RBCs, they go through a number of stages. The most important stages are the last two: when the cell is a **normoblast** and then a **reticulocyte**. The last stage during which the cell actually has a nucleus is the normoblast stage. After this stage, the nucleus disintegrates and the cell becomes a reticulocyte. While a small number of reticulocytes in the blood's circulation is normal, too many (in addition to the presence of normoblasts) could indicate that not enough mature RBCs are available to transport oxygen. Once again, this could be due to an injury, such as a hemorrhage, or a disease.

In order for these stem cells to mature into RBCs, they need a significant amount of nutrients, such as protein and iron. In fact, protein and iron are necessary in order for hemoglobin to synthesize. In addition, vitamins such as folic acid and B_{12} are needed in order for the stem cells' genetic material to synthesize in the bone marrow. Two necessary chemical agents, called factors, must be present in order for the stem cells to mature into RBCs: the **extrinsic factor** and the **intrinsic factor**. The source for the extrinsic factor, also known as vitamin B_{12}, is, as the name implies, external—food. The intrinsic factor comes from certain cells, called parietal cells, of the stomach lining. This factor then combines with food's vitamin B_{12} resources in order to prevent the vitamin's ingestion in the stomach so it can instead be absorbed in the small intestine. If the body is not getting a sufficient supply of the intrinsic or extrinsic factors, then the person might develop **anemia**.

Once the RBCs are produced, they live for approximately 120 days; past this time, they lose their durability and become fragile. At that point, they are removed from the circulatory system by the tissue macrophage system. Macrophages—which are contained not only in the bone marrow, but also in the liver and the spleen—are the lymphatic system's consumers. These old and failing RBCs are eaten (phagocytized) and digested by the macrophages. However, the iron in the RBCs is extracted and placed in the blood, eventually returning back to the bone marrow to be synthesized into new hemoglobin. This recycling process is done repeatedly. All excessive amounts of iron are stored in the liver until needed by the bone marrow.

But not every aspect of the RBCs can be recycled. As the macrophages are processing the RBCs, they take the "heme" portion of the hemoglobin and convert it into bilirubin. The liver then takes the bilirubin and excretes it into bile. The bile leaves the liver, travels to the small intestine and colon, and eventually leaves the body as the waste product known as feces. A small amount remains in the bloodstream and is responsible for the coloring of the urine. If bilirubin does not leave the body, it stays in the blood and can be a sign that the person is suffering from hepatitis or some other blood-related illness.

Blood Types

There are two kinds of red blood cell types important to the lymphatic system because they involve antigens and antibodies: the **ABO group** and the **Rh factor** (see Table 6.4 and Figure 6.7). The ABO group includes A, B,

TABLE 6.4
Blood Types: The ABO Group

Blood type	What are the antigens on the RBCs?	What are the antibodies present in the plasma?
A	A	Anti-B
B	B	Anti-B
AB	Both A and B	Neither anti-A nor anti-B
O	Neither A nor B	Both anti-A and anti-B

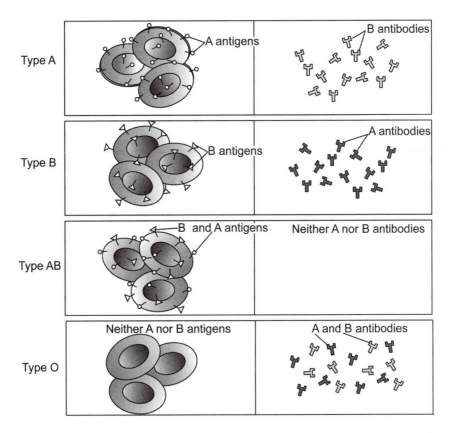

Figure 6.7   Blood Types and Antibodies. The column on the left shows red blood cells, and the right column shows plasma. The ABO blood types on the left include antigens, while the plasma includes the presence of antibodies. (Sandy Windelspecht/Ricochet Productions)

AB, and O blood types. A and B represent the presence of antigens on the RBC membrane. For example, there are A antigens on the RBCs of the patient with type A blood, while there are B antigens on the RBCs of the patient with type B blood. In someone with type AB blood, both antigens are present in the blood, while type O indicates that the person's blood contains neither A nor B blood.

The plasma of each person's blood contains naturally occurring antibodies for those antigens that are not present in the RBCs. This means that a person with type A blood has anti-B antibodies present in his plasma. In addition, the person with type B blood has anti-A antibodies in his plasma, and a type AB blood classification means that the person has neither A nor B antibodies in his plasma. The type O blood patient will have both anti-A and anti-B antibodies.

Blood-typing is extremely important when a blood transfusion is necessary for an operation or other medical procedure. Ideally, a patient should receive only a transfusion of their own blood type, or the procedure will not be successful. For example, if a type A patient needs blood and receives type B blood, then the patient's anti-B antibodies will bind to the donated blood's type B antigens. After the antibodies and the antigens are bound together, they would clump (**agglutination**) and then burst (**hemolysis**), which would defeat the entire purpose of the transfusion. In more serious circumstances, the RBCs that have ruptured would emit free hemoglobin, which would then clog and block the kidney's capillaries, eventually leading to renal failure or even damage. Because the type O patients have neither the A nor the B antigens, they are often considered "universal donors" who will not cause a reaction in the recipient.

Another important characteristic of RBCs is an antigen known as the Rh factor. While people with the Rh factor are considered Rh positive, those without the factor are called Rh negative, and their bodies do not have natural antibodies to this antigen. Therefore, during a blood transfusion, an Rh-negative patient should receive Rh-negative blood, while the Rh-positive patients should receive Rh-positive blood. If for some reason, an Rh-negative patient receives Rh-positive blood, the body will perceive the Rh factor as foreign, and therefore will then begin producing antibodies during that initial exposure. While there likely will not be a problem following this initial transfusion, subsequent exposures to Rh-positive blood when the anti-Rh-factor antibodies are already present could lead to hemolysis and potentially damage the kidneys.

Thymus

In addition to the bone marrow, the second primary lymphatic organ is the thymus, which is located under the sternum in an adult. In a fetus and an infant, however, the thymus gland is located below the thyroid gland, which is an endocrine gland below the larynx. As the body develops and grows, the thymus actually shrinks and becomes fat tissue, leaving only a small amount of the thymus in adults. The thymus reaches maximum size during puberty.

The T lymphocytes or T cells that are vital in the body to prepare the immune system to perform its primary duties are produced in the thymus. In fact, the term "T cell" is derived from "thymus-dependent cells." The hormone released by the thymic gland prepares the T cells to recognize antigens and other foreign invaders in the body and subsequently provide immunity.

During the fetal and infant stages, the immune system is immature, which is why babies are more vulnerable to disease and illness in comparison to children and adults. Early in life, the T cells begin (and the lymphocytes perpetuate) this protection as the immune system develops and matures. When a child is two years old, his or her immune system is typically considered mature and fully functional. While infants are routinely given vaccines to boost their immune systems, some vaccines, such as the measles vaccine, are not recommended for children younger than 18 months, because many medical professionals believe that the child's immune system would not be strong enough to respond properly, thus putting the child in danger of not getting the full benefit of the vaccine, or even having a reaction.

Secondary Lymphatic Organs

In addition to the lymph nodes, thymus, and the bone marrow, the lymphatic system functions with help from five other important organs: the spleen, tonsils, adenoid, Peyer's patches, and appendix. This section will describe how these secondary lymphatic organs function and help to protect the body from microbes and other foreign antigens that could lead to illness or cause various kinds of diseases.

Spleen

The spleen is protected from harm by the lower rib cage, which encases the organ behind the stomach and is inferior to the diaphragm. While in the fetal stage, the spleen produces RBCs, although this process is taken over by the bone marrow shortly after birth.

There are three primary operations performed by the spleen following birth. One of the spleen's functions is to produce lymphocytes, which then enter the blood and serve as one of the primary tools for the immune system to fight off antigens. Secondly, the spleen contains plasma cells, which produce antibodies that also ward off foreign antigens and microbes. Finally, the spleen also contains macrophages which have the ability to consume, or phagocytize, foreign materials floating around in the blood. In addition, the spleen's supply of macrophages serves to destroy old RBCs and produce bilirubin, which is eventually extracted to the liver and excreted from the body as bile.

Because of its two-part composition, this organ is often described as two organs. One portion of the spleen is composed of lymphatic sheaths and germinal centers called white pulp. The second portion is known as red pulp, and consists of macrophages. The white pulp's function is considered immune, while the red pulp's function is considered phagocytic. The white pulp is in charge of producing the antibodies, and this region is also where B and T cells, in addition to plasma cells, are produced and mature. The red pulp is kind of a cleaning machine; it removes unwanted matter, such as bacteria. In addition, the red pulp also acts as a reservoir for other lymphatic elements such as white blood cells and platelets.

Doctors do not consider the spleen a "vital organ," because it performs the same functions as some other organs. For instance, the liver and bone marrow can remove RBCs from the circulatory system, and the lymph nodes will produce lymphocytes and monocytes, in addition to destroying pathogens. However, doctors and researchers have found that without a spleen, a person is more vulnerable to certain bacterial infections such as pneumonia and meningitis.

Tonsils, Adenoid, and the Peyer's Patches

Tonsils are a type of lymph nodule found in the throat's pharynx. Tonsils are oval-shaped, pink masses of lymphatic tissue. There are three kinds of tonsils named for their location in the pharynx. Along the lateral walls of the pharynx are the palantine tonsils, while the adenoid, or the pharyngeal tonsil, are located on the posterior wall. The lingual tonsils are located on the base of the tongue. The tonsils and adenoid are composed of lymphatic tissue, just like the lymph nodes located in the neck, groin, and armpits.

The adenoid is a single mass of tissue; therefore, it is incorrect to refer to it as "adenoids." As explained above, it is located in the upper part of the throat, behind the nose and above a part of the throat called the uvula. This area of the throat is called the nasopharynx. While the tonsils can be seen simply by opening the mouth wide, the adenoid can be viewed only through the use of special mirrors and instruments that are passed through the nose by a doctor or other medical professional.

The tonsils form a kind of ring of lymphatic tissue around the pharynx. This is a key location because it is near the entrance to breathing passages, in addition to being where food first enters the mouth. Therefore, the tissue can capture germs and pathogens that are coming into the body through food and air, and it can act as a sort of filter for the lymphatic system. This function is especially important during the initial years of life, but becomes less important as the body and lymphatic system matures. Children who suffer frequent infections of the tonsils (called tonsillitis) may have to have their palatine tonsils and their adenoid removed. Some signs of infected tonsils might be noisy breathing, snoring, difficulty swallowing (especially solid foods), and choking or gasping while sleeping. The surgery to remove the tonsils is known as a tonsillectomy. It is important to note that when children have their tonsils removed, they do not suffer immunity loss, because the body has lined up redundant systems, such as the other lymph nodules, that will serve the same function if the tonsils are surgically removed.

In the small intestine, there is an abundance of lymphatic tissue in order to filter out pathogens that might be brought into the body through eating and drinking. One type of lymph nodule grouping, located in the small intestine, is called Peyer's patches. In the small intestine, the Peyer's patches work to remove pathogens that are invading the body through the digestive system. There are also single lymph nodules, known as solitary lymph nodules, located in the lower part of the small intestine (see Chapter 3 on the digestive system for more information on how the body breaks down and processes food and other substances).

Appendix

A small, blunt-ended tube, the appendix is considered part of the large intestine, even though its walls are rich with lymphatic tissue. However, it is recognized that the appendix is not actually part of the digestive tract because of differences in its tissues from the tissues of the small and large intestine. In some cases, fecal matter or waste can become impacted in the appendix, which can cause it to become inflamed, a condition known as **appendicitis**. If this occurs, the appendix will be surgically removed, a process known as an **appendectomy**.

The Immune Response: Cell-Mediated and Antibody-Mediated Responses

In order to maintain good health and homeostasis, the human body must continually protect itself against harmful disease-causing substances. This section will explore how the lymphatic system is able to protect the body by providing **immunity**, the ability to fight off certain infectious substances that can lead to disease and illness. It takes various elements of the lymphatic system, including lymphocytes and antibodies, which provide a strong defense to counter these disease-producing pathogens.

Before each of these lymphatic tools is explained, however, it is important to distinguish between nonspecific and specific defenses. Nonspecific defenses are reactions that involve a variety of pathogens and microbes. These defense reactions occur in the skin, mucous membranes (such as nasal passages), the stomach, and the respiratory tract. For instance, lysozyme is an enzyme produced in the eyes' lacrimal glands as well as the glands in the mucous membranes of the nose and mouth; it is able to destroy harmful microbes, and thus it is considered one of the body's nonspecific defenses. Also, stomach lining is able to secrete an appropriate amount of hydrochloric acid in order to kill harmful microbes present in food, although some food-borne pathogens can still be harmful. When one breathes, it is natural to inhale dust particles that then settle in the respiratory tract. Microbes attach themselves to these particles, but thread-like structures called cilia move these microbes up through the respiratory tract to be coughed up or spat out.

While the body's nonspecific defense system is effective against microbes and some pathogens, it needs help to fight, especially against certain toxins produced by pathogens. Therefore, the body is equipped with a second line of defense, known as the specific defenses. The specific defenses involve the production of antibodies, which serve to inactivate substances called antigens (pathogens and their related toxins). Antigens act as chemical markers that identify cells as invasive substances. Human cells have their own antigens, which recognize foreign antigens as dangerous and are subsequently destroyed, thus activating an immune response. Examples of foreign antigens include bacteria, viruses, fungi, protozoa, and **malignant** cells, which are abnormal cells such as those associated with cancer. The response that results from the antibody-antigen reaction is specific. Only a specific antibody can fight off a particular antigen (this relationship will be explored later in this chapter).

Nonspecific and specific immune responses are also referred to as innate and adaptive responses, and both systems work together to identify harmful invaders to the body, and then contain and eliminate them. The innate or nonspecific system is always on alert, and is prepared to react to any and all invaders. While their action is rapid, the innate immune response is also limited, but keeps the harmful pathogens from invading the body to a significant degree. However, the adaptive immune system can come in and is equipped with the powerful, specified tools to completely eliminate the pathogen.

Thymus Gland

It is important to emphasize the vital role that the thymus gland plays in immunity. One of the lymphatic system's primary organs, scientists and doctors noted in the early years of immune research that children born without this organ could not fight off infection. Research has indicated that the thymus gland is instrumental in structuring and organizing the body's lymphatic system from the fetal years through the initial years after birth. The primary function of the thymus gland is to prepare lymphocytes to participate in the immune response.

Lymphocytes

The lymphatic system's organs—the lymph nodes, thymus gland, spleen, and bone marrow—all contain lymphoid tissue, which is home to two kinds of lymphocytes that each respond to antigens in different ways: T cells and B cells.

In the embryo stage, T cells are produced in the bone marrow and thymus. While passing through the thymus, they mature with the help of the thymic hormones. Scientists believe that the thymus gland alters the lymphocyte's DNA so they become T cells. These cells then travel to the spleen, lymph nodes, and lymph nodules, where they then are produced following birth. After the initial production, the T cells circulate through the body's blood network, and then lodge in the lymph nodes and other lymphoid tissue. T cells are small lymphocytes that attach to antigens. Once attached, the T cells secrete certain enzymes that dissolve the antigen's membrane and digest its contents, which destroy both the antigen and the T cell. These lymphocytes primarily kill antigens produced by fungus cells, viruses, and bacteria that result from slow-developing infections and diseases. T cells are also responsible for the rejection that can result from an organ transplant.

The second kind of lymphocyte, the B cell, is produced in the bone marrow during the embryonic stage, although it soon moves into the spleen and lymph nodes and nodules. When B cells come into contact with a specific antigen, they become plasma cells, which produce antibodies that are released into the blood's circulation. Once these specified B cells are produced after coming into contact with a specific antigen, they can remain in the lymph nodes for years, on alert to attack if the antigen is once again introduced into the body.

Antibodies

As stated earlier, antibodies are proteins that are produced in response to foreign antigens. Also called **immunoglobulins** or gamma globulins, it is important to note that antibodies in themselves do not destroy foreign antigens. Instead, antibodies attach themselves to foreign antigens, marking these substances so the body knows to destroy them.

Also as stated earlier, there is one specific antibody for one specific antigen, which means that if the need occurs, the immune system has the capacity to respond to millions of antigens by producing millions of different antigen-specific antibodies. These millions of antibodies are separated into five classes: IgG, IgA, IgM, IgD, and IgE.

The Immune Response

One of the main goals of the lymphatic system's immune response is to destroy a harmful pathogen, and the first step to achieve this goal is for the body to recognize the antigen associated with this pathogen as foreign. While both T cells and B cells can provide this recognition function, the immune response is more effective if the antigen is dealt with by macrophages and helper T cells, which are a specialized group of T lymphocytes. This foreign antigen is first consumed, or phagocytized, by a macrophage. Parts of this antigen then become attached to the macrophage's cell membrane. Also located on this cell membrane are "self" antigens, which are the safe type of antigens found in other cells throughout the individual. When the helper T cells come into contact with this macrophage, it will detect not only the "self" (and harmless) antigens, but also the foreign antigens. This helper T cell is then on alert and sensitized to this macrophage that contains parts of the foreign antigen.

Cell-Mediated and Humoral Immunity

There are two types of specific (or adaptive) immunity: cell-mediated and humoral. T cells are responsible for cellular or cell-mediated immunity, which refers to immune response in which macrophages and T cells participate, while humoral immunity involves B cells, T cells, and macrophages. While T cells are associated with cellular immunity, B cells are responsible for humoral immunity, because the B cells are providing protection as they circulate through the blood and tissues of the body. Humoral immunity protects against more acute diseases than are warded off by T cells, such as pneumonia, staphylococcal infection (staph infection), and streptococcal infection, which is also known as strep throat.

Figure 6.8 depicts the cell-mediated immunity process. This process does not produce antibodies, although it is successful against intracellular pathogens, including viruses, fungi, malignant cells, and foreign tissue grafts. The initial step in this process is to activate the T cells, which

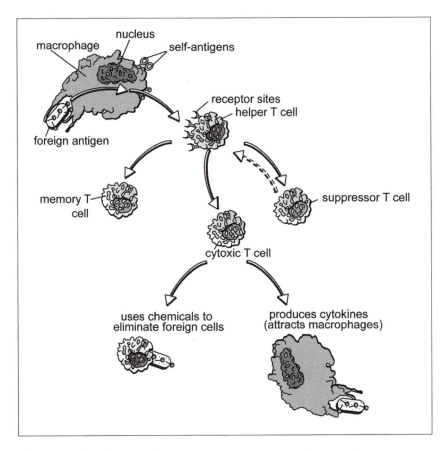

Figure 6.8 Cell-mediated immune response. (Sandy Windelspecht/ Ricochet Productions)

occurs when the helper T cells and macrophages recognize a foreign antigen. Now, recall that these activated T cells are antigen-specific, meaning individual T cells are successful only against certain antigens. But when these T cells are activated, they divide numerous times into two kinds of cells: memory T cells and cytotoxic (killer) T cells.

While the memory T cells can always recall a specific antigen and become sensitized upon its presence in the body, the cytotoxic cells are able to eliminate these foreign antigens by destroying their cell membranes. These cells also produce a certain chemical called a **cytokine** that attracts macrophages to the area where an antigen is present, and then activates them to consume, or phagocytize, the antigen. The effect of these two types of T cells working together ensures that a harmful antigen, such as a virus, is quickly detected, destroyed, and then prevented from reproducing other virus-infected cells in the body.

Another kind of cell, called the suppressor T cell, is also present in this immune system. Once the memory and cytotoxic cells work to eliminate the antigen, the suppressor cells work to stop the immune response once the antigen has been destroyed. However, if the antigen reappears, the memory cells will initiate the immune response.

Unlike the cell-mediated immunity, humoral immunity does result in antibody production (Figure 6.9). Once again, the initial step in this immune response is also the recognition of the foreign antigen; the helper T cells, B cells, and macrophages are all involved. The helper T cells recognize the antigen and then alert the B cells, which activate other B cells that might be specialized or specific in combating the antigen. These sensitized B cells go through numerous divisions, which results in the production of two types of cells: memory B cells and plasma B cells. The memory B cells will remember the antigen, while the plasma B cells will produce specific antibodies against this one invading antigen.

After these antibodies are produced, they bond to the antigen, which forms an antigen-antibody complex. This complex is then marked, or undergoes **opsonization**, which means that macrophages or neutrophils will know that this antigen must be phagocytized. The creation of this antigen-antibody complex also begins the process of **complement fixation**. A complement is a family of about 20 different plasma proteins. These proteins circulate through the body's blood network until they are activated, or "fixed," by the formation of the antigen-antibody complex. This fixation process can be complete and thorough, but it can also be partial. Complete fixation is successful if the antigen is cellular, which is often the case with bacterial antigens. In this instance, the complement proteins will bond to the complex and to each other, surrounding the antigen with an enzymatic structure that ultimately inflicts damage on the cells to a destructive degree, thus killing the cell.

But only partial complement fixation takes place if the antigen is not cellular, which would occur if the antigen were a virus. In this instance,

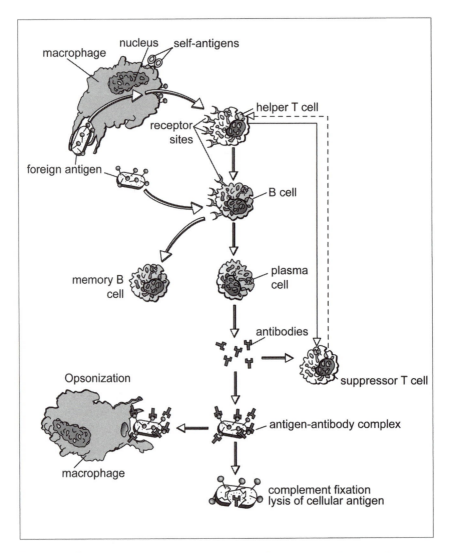

Figure 6.9   Humoral immune response. (Sandy Windelspecht/Ricochet Productions)

only some of the complement proteins bond to the antigen-antibody complex, thus prompting the antigen to go under **chemotaxis**, which is another labeling mechanism that attracts macrophages to phagocytize the antigen.

As stated earlier, once the antigen has been eliminated, the suppressor T cells step in and work to halt the immune response. This is vital to the health of the lymphatic system so that the body doesn't overproduce antibodies, which could trigger an autoimmune response.

Innate Immunity

Two important types of innate, or nonspecific, defenses are the inflammation response and the release of interferons. The inflammation of body tissue is a nonspecific response to tissue injury or pathogen invasion. The inflammatory response has three primary goals: the isolation and elimination of the harmful pathogens, the removal of debris from the injury site, and the preparation of the injury site for healing and repair.

When a pathogen, such as bacteria, enters the skin by breaking through the external skin wall, the macrophages present in that region of the skin promptly descend and go to work phagocytizing the microbes (Figure 6.10). As the initial line of defense, the macrophages fight infection to some extent, although they are not able to shoulder the work on their own. In fact, they are relatively stationary by nature, although they can travel to other sites near that initial region of invasion if necessary. As the macrophages go to work on the microbes, mast cells in the area of tissue damage release histamine, which prompts **vasodilation**.

As a result of vasodilation, blood vessels expand, delivering an increased amount of blood, phagocytic leukocytes, and plasma proteins to the injury site. In addition, the release of histamine causes capillaries to become more permeable, which causes a capillary's pores to enlarge. When this occurs, plasma proteins that are normally trapped in the capillaries are able to escape and travel to the inflamed tissue.

The arrival of the leukocytes and plasma proteins, along with the accompanying increased amounts of blood, cause fluid to build up in the injured tissue area. This leads to swelling, one characteristic of inflammation, in addition to redness and heat, which are due to the increased flow of warm, arterial blood to the region. These substances also sensitize the afferent neurons in the area, which causes the feelings of pain and tenderness.

In addition to inflammation, interferons are another nonspecific immune defense that is important against viral infections. Interferons are

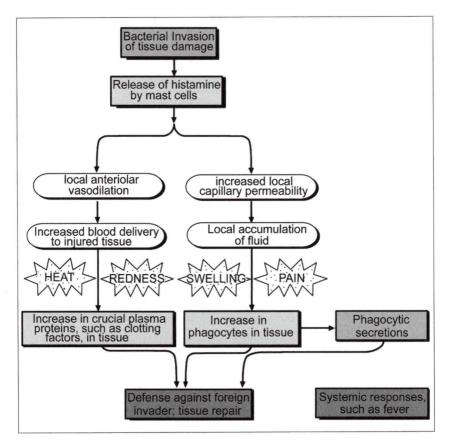

Figure 6.10   Inflammation and the Body's Immune Response.
The production of inflammation as part of the body's immune response.
(Sandy Windelspecht/Ricochet Productions)

not one single chemical or molecule, but rather a family of proteins that interfere, albeit briefly, with a virus's ability to reproduce in other cells. These proteins derive their name from the ability to interfere with viral replication. When a virus invades a cell, the cell's genetic components prompt it to produce an interferon, which is then secreted into the extracellular fluid. In this way, the interferon is able to warn nearby healthy cells of the viral presence, thus helping them to prepare for resistance.

Once the interferon is in the extracellular fluid, it attaches itself to receptors present on the plasma membranes of these neighboring cells, and even distant cells. While it is important to note that interferons themselves do not have antiviral capabilities, these proteins prompt potential host cells to begin production of virus-blocking enzymes, which can break down the virus's genetic components (including protein synthesis) that are vital for replication. These antiviral enzymes are dormant unless a virus invades their host cells, in which case the enzymes detect the virus's genetic material (nucleic acid) and go to work. However, this activation can occur only during a limited amount of time, and therefore, it is only a temporary defense against viral infections. After this time, specified immune responses are needed, such as those that relate to antibody production.

In addition to their power against viral replication, interferons also help to boost other immune-related mechanisms, including enhancing the phagocytic behavior of macrophages and the production of antibodies. Because they have been shown to slow cell division and tumor growth, interferons also have anticancer benefits, although the extent of these benefits is still being researched.

Genetic and Acquired Immunity

The human body's immune system is divided into two categories: **genetic immunity** and **acquired immunity**. Genetic immunity is determined by DNA and therefore does not involve antibodies; rather, it is the immune system that we are born with as part of our genetic composition. This is specific to a species, however, which means that some species have immunity to pathogens that are harmful to other species. For example, while the measles virus is harmful to humans, it does not affect dogs and cats because of their genetic immunity to the pathogens associated with the virus. Certain viruses that are dangerous to plants are harmless to humans because our genetic makeup protects our cells and tissues from these pathogens (see Table 6.5).

Acquired immunity involves antibody production. Within this category, there are two subcategories: passive and active immunity. While active or

TABLE 6.5
Genetic and Acquired Immunity

Type of immunity	Definition	Examples
Genetic	Part of the human body's genetic makeup (DNA); does not involve antibodies; different from other species	
Acquired	Involves antibody production	
	Passive (Natural)	Transmission of antibodies through placenta from mother to fetus; antibodies from other resources; antibody transmission through breast milk from mother to baby
	Passive (Artificial)	Antibody injection in the form of gamma globulins or immune globulins after prior exposure
	Active (Natural)	Production of own antibodies; disease recovery through antibody and memory cell production
	Active (Artificial)	Antibody and memory cell production as a result of a vaccine

"self-generated" immunity involves the production of antibodies, passive or "borrowed" immunity is when antibodies are acquired through the direct transfer of antibodies produced by another person or animal. An example of this transfer includes the IgG antibody movement from mother to fetus through the womb's placenta. Another example includes breast-feeding, when a mother's milk is enriched with IgA antibodies. This is important because even though these acquired antibodies are broken down within a month, they provide that initial protection against infections while the newborn begins to develop his own immune response system, because the production of antibodies does not happen until at least one month after birth. (See Table 6.6 for a breakdown of passive and active immunity.)

In some cases, passive immunity is used in response to a lethal toxin or dangerous viral-related infectious agent, such as when someone has

TABLE 6.6
Active and Passive Immunity

	Active immunity	Passive immunity
Antigen exposure	The immune system is exposed to the antigen either naturally as a result of an infection, or artificially as a result of a vaccination.	No antigen exposure is required.
Antibody source	Antibodies are produced by the body's immune system upon exposure to an antigen.	Preformed antibodies are borrowed from another source and introduced into the immune system.
Administration of antibodies	Injection or attenuated antigen is needed for artificial active immunity.	For artificial passive immunity, patient is injected with borrowed antibodies.
Develop time for resistance	Antibody production can occur in several weeks or days.	Antibody production is immediate.
Resistance time	Sometimes lifelong.	Only a few weeks.

been bitten by a dog afflicted with the rabies virus or a snake with poisonous venom. When this occurs, the injected antibodies have been harvested from another, often nonhuman source that has produced antibodies following exposure to the antigen in an attenuated form, which means a less potent or virulent form of the antigen. Some of the most common animals used for passive immunity procedures are horses and sheep. Because the antibody injection (or serum) is a foreign substance, the body's immune system might launch an attack in response, which would result in an allergic reaction.

Vaccines and Immunity

While active immunity is the production of one's own antibodies, it may be done naturally, or even stimulated through artificial means. Once a person has recovered from a disease, such as chickenpox, his body now has specific antibodies and memory cells that will activate if that disease's pathogens enter the body again. This is called naturally acquired active immunity. Vaccines help bring about artificially acquired active immunity, because the mechanism stimulates the production of memory cells in addition to antibodies against a disease.

The vaccine mechanism was invented by Edward Jenner, a British physician in 1797, when he found that patients could become immune to the devastating effects of the smallpox disease (which had a mortality rate of 40 percent) through exposing or inoculating patients with small amounts of the cowpox disease, which is a weaker form of the smallpox disease.

The goal of vaccination, which begins soon after birth for infants, is to equip the body with active immunity by creating a memory system that calls on B and T cells to defend the body when recognizable disease-causing agents enter the blood and lymphatic system. Usually injected through a needle, a vaccine contains a weakened, or attenuated, form of a specific disease-causing germ. In some cases, the vaccine will contain an inactivated form of the germ, which will produce the toxins associated with the disease upon entering the bloodstream.

Once in the body, the immune system produces antibodies against these germs. Because these germs are weakened or dead, they are often not strong enough to make the patient sick, but just strong enough to stimulate the immune system to produce antibodies. It is important to note, however, that just like any medical procedure, there are risks associated with vaccines. Some people have an allergic or other adverse reaction to certain vaccines, although the chances are minimal. Once these antibodies are floating around in the body, memory cells are formed. Both antibodies and these memory cells are then on alert in case the disease-related germ or pathogen invades the body at a later date.

Vaccines come in different forms. Some are used in combinations, such as the DTP (diphtheria, tetanus, pertussis) and the MMR (measles, mumps, rubella) vaccines that children receive during their early years to boost their immune systems. Vaccines can be produced from three types of microbe-related materials: inactivated (killed), attenuated (live), and synthetic (produced in a laboratory). In the first type, inactivated vaccines are produced by killing the disease-causing microbe through chemical means, making them stable. Most inactivated vaccines stimulate a rather weak immune response in patients; therefore, the vaccines must be given several times. Examples of inactivated vaccines include those for cholera and hepatitis A.

Live and synthetic vaccines are produced in a laboratory setting in order to eliminate its viral and disease-causing characteristics. Unlike inactivated forms, these vaccines produce both cell-mediated and antibody-mediated immunity, and usually require only one dose. Examples of these vaccines include those for yellow fever, measles, rubella, and mumps.

Another kind of vaccine formulation is a **toxoid**. A toxoid contains an inactivated form of a toxin, which is the harmful substance produced by a microbe. In general, microbes are not dangerous or disease-causing, although some toxins emitted by microbes can cause illness. For example, in a normal environmental setting with plenty of oxygen, the bacterium associated with tetanus is harmless. But when this bacterium is in an environment without oxygen, the bacterium produces a poisonous toxin. These potent toxins are treated with materials, such as a sterile water and formaldehyde solution called formalin, which inactivates these toxins. Diphtheria and tetanus vaccines use toxoids.

Summary

One of the human body systems' most important—and difficult—jobs is to protect the body against the myriad of organisms that threaten its health and homeostasis. While each of the human body systems has its own defense mechanisms, the duty to protect the body against dangerous invaders falls primarily to the lymphatic system. This system contains specialized cells called lymphocytes that detect threatening organisms and put into motion an immune response that eliminates them from the body. The immune response also protects the rest of the body against free pathogens and cells that might already have been infected. In addition, this response mechanism actually remembers the infection for future defense purposes; if the invading organism enters the body again, protection will be in place.

Unlike other systems in the human body, there are only a few primary and secondary organs in the lymphatic system. The bone marrow and thymus are the primary organs, while the secondary organs include the spleen, tonsils, adenoid, Peyer's patches, and appendix.

Glossary

Abduction Withdrawal of a part of the body from the body's axis.

ABO group The name of the genetic system that determines human blood groups. Named for the presence of A and B carbohydrates on the surface of the cell, or the absence of the carbohydrates in the case of the O group. This system uses four possible combinations: A, B, AB, or O.

Acetylcholine (ACh) A neurotransmitter released in the central and peripheral nervous system, specifically at neuromuscular joints.

Acetyl Coenzyme A (acetyl CoA) A molecule that enters the citric acid cycle to produce energy. The acetyl CoA can come from sugars that have gone through glycolysis, or it can come directly from fats or proteins in the cell.

Acidosis An abnormal increase in the acidity of the body's fluids.

Acquired Immunity A type of immunity that is not the result of genetic inheritance, but rather due to the exposure to some antigen and the resulting response by the immune system.

Actin One of the major proteins involved in muscle contraction. Actin proteins form a long fiber within the muscle contractile unit. Myosin, the other major protein involved in muscle contraction, attaches to the actin filaments and pulls the muscle shorter.

Action Potential A change in the electrical charge of a nerve cell following the transmission of a nerve impulse.

Adaptation The state of sensory acclimation in which the sensory awareness diminishes despite the continuation of the stimulus.

Adduction Movement of a limb toward the median line of the body.

Adenosine Diphosphate (ADP) A chemical substance produced through digestion and used in cell respiration and energy production.

Adenosine Triphosphate (ATP) A chemical substance produced from aerobic cell respiration that is the muscle's direct source of energy for movement.

Adipocytes Cells that have large holes filled with fat.

Adipose Capsule The central layer surrounding the kidney, composed of fatty tissue.

Adrenal Glands The hormone-releasing glands located above the kidneys.

Adrenaline Also known as epinephrine, a hormone produced by the adrenal glands that helps regulate the sympathetic division of the autonomic nervous system. During times of stress or fear, the body produces additional amounts of adrenaline into the bloodstream, causing an increase in blood pressure and cardiac activity.

Adrenocorticotropic Hormone (ACTH) Hormone produced by the pituitary gland, which stimulates the release of hormones from the adrenal cortex. Also called corticotropin.

Aerobic Anything having to do with acquiring oxygen from the air.

Aerobic Cell Respiration A chemical process that allows the cell to produce energy from glucose and oxygen.

Aerobic Exercise Exercise in which energy is made by processes involving oxygen. Types of aerobic exercise include swimming, biking, and jogging.

Afferent Nerves Fibers coming to the central nervous system from the muscles, joints, skin, or internal organs.

Afferent Vessels A form of vessel that brings fluid towards an organ or lymph node.

After-image An image of a visual nature that exists even after the visual stimulus has ceased.

Agglutination The clumping of blood. This can occur if a patient with a certain blood type is given blood of another type.

Agonists Hormones that bind to their receptor and elicit a specific biological response.

Albumin The most abundant plasma protein. It makes up 55 percent of the total protein content of plasma. It is involved in maintaining blood volume and water concentration.

Aldosterone A hormone secreted by the adrenal glands in the kidneys that increases sodium reabsorption.

Alkaline A term used to indicate a pH of 7 or greater. Sometimes also called basic.

Allele A variation of a gene that encodes for a specific trait. It is usually due to minor variations in the DNA at the molecular level.

Allergies Hypersensitive reaction to a particular substance or allergen; symptoms vary in intensity.

Alveoli Tiny air sacs in the lungs. They exchange oxygen and carbon dioxide between the lungs and the blood.

Amino Acids The building blocks of proteins.

Amphiarthrosis Joint that permits only slight movement.

Amphiphatic Molecules A term given to a molecule that has both hydrophilic and hydrophobic properties.

Anaerobic Exercise Exercise in which energy is made by processes that do not involve oxygen. Types of anaerobic exercise include weight lifting and sprinting.

Androgens Male sex hormones produced by the gonads and adrenal cortex.

Anemia A reduction in the number of red blood cells in the body, resulting in an insufficient number of hemoglobin molecules to carry oxygen to the tissues of the body. This may result in tissue color changes, weakness, and increased susceptibility to disease.

Anions Negatively charged particles.

Annulus Fibrosus A ring of fibrous connective tissue that serves as an anchor for the heart muscle and as an almost continuous electrical barrier between the atria and ventricles.

Antagonistic Muscle Pair Two muscles that have an opposite action, such as the muscle that bends the arm and the muscle that straightens the arm. The antagonistic pair controls and stabilizes the elbow as it bends and straightens.

Antagonists Hormones that bind to the receptor but do not trigger a biological response. By occupying the receptor, an antagonist blocks an agonist from binding and thus prevents the triggering of the desired effect within the cell.

Anterior Situated in front; at or toward the head end of a person or animal.

Anterior Pituitary The lobe of the pituitary that secretes hormones that stimulate the adrenal glands, thyroid gland, ovaries, and testes.

Antibody Proteins that attack antigens.

Antidiuretic Hormone (ADH) Hormone produced by the pituitary gland that increases the permeability of the kidney ducts to return more fluid to the bloodstream. Also called vasopressin.

Antigens Invading organisms and materials that enter the human body. The body may mount a defense with antibodies.

Antioxidants Compounds that prevent oxidative damage to organic molecules. Vitamins C and E are examples of antioxidant nutrients, as is the mineral selenium.

Aorta The largest artery in the human body. It supplies oxygenated blood from the left ventricle of the heart to the branching arteries, which in turn supply oxygen to all parts of the body.

Aortic Arch A large, rounded section of the aorta that occurs above the heart, just after the aorta leaves the right ventricle.

Aortic Bodies Chemoreceptors found in the aortic arch, the curved portion between the ascending and descending parts of the aorta.

Appendectomy The surgical procedure that is used to remove an inflamed,
diseased, or ruptured appendix.

Appendicitis An inflammation of the appendix. This is usually caused by an infection of the appendix and results in fever, pain, and loss of appetite.

Appendicular Skeleton The skeletal structures composing and supporting the appendages; these include the bones of the shoulder and hip girdles as well as those of the arms and legs.

Arterial Baroreceptor Reflex The mechanism that provides oversight and maintenance of the blood flow by responding to slight changes in blood pressure.

Arterial System The portion of the circulatory system that delivers oxygen-rich blood to the body tissues.

Arteries Larger blood vessels that deliver oxygen-rich blood to the body tissues.

Arterioles Smaller blood vessels that deliver oxygen-rich blood to the body tissues.

Arteriovenous Anastomoses Blood vessels that directly connect arterioles to venules. Commonly, blood travels from arterioles to capillaries to venules. Arteriovenous anastomoses are typically found in only a few tissues.

Arthritis An inflammation of the joints.

Asexual Reproduction Reproduction in which genetically identical offspring are produced from a single parent.

Atherosclerosis Also known as hardening of the arteries. It is a narrowing of arterial walls caused by deposits, collectively called plaque, that create rough, irregular surfaces prone to blood clots.

Atria Plural of atrium.

Atrium In the human heart, it is one of the heart's two upper chambers. The plural form is atria.

Autocatalytic Process A chemical reaction in which the products of the reaction are responsible for initiating the start of the reaction.

Autocrine The action of a hormone on the cells that produced it.

Autoimmune A term used to describe an immune response to the patient's own body. An autoimmune disease is therefore one that attacks part of the patient's body.

Autoimmune Disease This occurs when the immune system incorrectly identifies the tissues of the body as foreign material, and begins an immune response against the cells or tissue. Lupus and forms of diabetes may be caused by an autoimmune response.

Autonomic Nervous System The part of the nervous system that controls involuntary actions and rules the variations of the heart rate.

Axial Skeleton The central supporting portion of the skeleton, composed of the skull, vertebral column, ribs, and breastbone.

Axon A single nerve fiber that carries impulses away from the cell body and the dendrites.

B Cells Also known as B lymphocytes. They are one of two main types of lymphocyte, and participate in the body's immune response.

B Lymphocytes *See* B Cells.

Baroreceptors Pressure detectors located in the major arteries. Part of the arterial baroreceptor reflex, they sense a dip or spike in blood pressure.

Basilar Artery A blood vessel that arises from the vertebral arteries and joins with other cerebral arteries to form the circle of Willis.

Basophil A type of granulocyte that appears to be active in the inflammatory process.

Bayliss Myogenic Response The mechanism by which smooth muscle cells impart muscle tone to the blood vessels.

Bilirubin A waste product produced by the liver that is the result of the breakdown of red blood cells. It is released into the small intestine, but some is reabsorbed back into the blood and excreted with the urine.

Binucleate Cell A cell that contains two nuclei.

Bioavailability A term of nutritional analysis that indicates how much of a nutrient in a food is actually available to the body for absorption by the gastrointestinal tract.

Biomarker A molecular clue indicating the presence of disease or the genetic predisposition for disease.

Biomolecule A general classification for any of the four groups of organic molecules that are used in the building of cells—proteins, carbohydrates, lipids, and nucleic acids.

Bipedal An animal that walks on two feet.

Bladder A hollow, muscular organ that stores urine for elimination.

Blastocyst An embryonic stage following the morula stage characterized by outer trophoblast cells, an inner cell mass, and a central, fluid-filled cavity.

Blood The fluid that contains the plasma, blood cells, and proteins and carries oxygen, carbon dioxide, nutrients, waste products, and other molecules throughout the body.

Blood Cells Cells contained in the plasma of the blood. *See also* Red Blood Cells and White Blood Cells.

Blood Pressure The force of the blood against the walls of the blood vessels.

Blood Sugar Level The amount of glucose in the blood.

Blood Type A form of blood, determined by the presence or absence of chemical molecules on red blood cells. A person may have type A, B, O, or AB blood.

Blood Vessels Also known as the vasculature. These are the tubes of the circulatory system that transport the blood throughout the body.

Bohr Effect High concentrations of carbon dioxide and hydrogen ions in the capillaries in metabolically active tissue that decrease the affinity of hemoglobin for oxygen and leads to a shift to the right in the oxygen dissociation curve.

Bolus The name given to the mass of food that accumulates at the rear of the oral cavity for swallowing.

Bone Marrow The site in the body where the cells of the lymphatic system originate.

Bowman's Capsule The bulb surrounding the glomeruli. It provides an efficient transfer site for water and waste products to move from the blood to the urinary system.

Brachial Artery The blood vessel in the upper arm that accepts blood from the subclavian artery by way of the axillary artery, and travels down the arm to supply the ulnar, radial, and other arteries of the forearm.

Brachial Vein The blood vessel that collects blood from the ulnar vein and empties into the axillary vein.

Brachiocephalic Artery Also known as the innominate artery. This short blood vessel arises from the aortic arch, and branches into the right common carotid artery and the right subclavian artery.

Brachiocephalic Veins Also known as the innominate veins. This pair of veins arises from the convergence of the internal jugular and subclavian veins and flows into the superior vena cava.

Brain Other than the spinal cord, the primary organ in the nervous system.

Brain Stem This area of the brain connects the cerebrum with the spinal cord and is also the general term for the area between the thalamus and the spinal cord, which includes the medulla and pons.

Bronchi The two large air tubes leading from the trachea to the lungs that convey air to and from the lungs.

Bronchial Vein One of two main blood vessels that collect newly oxygenated blood from the bronchi and a portion of the lungs, and deliver it through one or more smaller veins to the superior vena cava.

Bronchiole Any of the smallest bronchial tubes that end in alveoli.

Buccal Cavity Another term commonly used to describe the oral cavity. It technically represents the space between the back of the teeth and gums to the rear of the mouth.

Bundle of His A thick, conductive tract located in the heart that transmits the electrical signal from the AV node to the Purkinje fibers in the base of the ventricle wall.

Bursa A sac of fluid within a joint.

Calcitonin Hormone produced by the thyroid gland that influences calcium and phosphorous levels in the blood.

Calcitrol A hormone secreted by the kidneys that increases the levels of calcium and phosphorous in the blood.

Calorie A measure of how much energy food contains.

Cancellous Bone Bone that has a latticework structure, such as the spongy tissue in the trabecular bone.

Capillaries The tiniest blood vessels. They are the sites of exchange: At body tissues, blood in the capillaries delivers oxygen and nutrients, and picks up carbon dioxide and waste products; and at the lungs, blood in the capillaries drops off carbon dioxide and picks up oxygen.

Carbon Monoxide Poisoning A medical condition arising when a person is exposed to carbon monoxide gas. Prolonged exposure can be fatal.

Cardiac (Heart) Muscle The type of muscle found in the heart.

Catecholamines A class of hormone (including epinephrine and norepinephrine) synthesized in the adrenal medulla that is involved in the body's stress response.

Cation A positively charged particle.

Cell Body Main mass of the neuron that contains the nucleus and organelles.

Central Nervous System Division of the nervous system that contains the brain and the spinal cord.

Cerebellum Located towards the back of the medulla and pons, this portion of the brain is in charge of many subconscious aspects of skeletal muscle functioning, such as coordination and muscle tone.

Cerebral Aqueduct The tunnel that runs through the midbrain, allowing cerebrospinal fluid to travel from the third to the fourth ventricle.

Cerebral Cortex This area of the brain is the gray matter located on the surface of the cerebral hemispheres. The cerebral cortex includes the brain's motor, sensory, auditory, visual, taste, olfactory, speech, and association areas.

Cerebrospinal Fluid (CSF) The fluid in the spinal cord's central canal that serves as the fluid for the central nervous system. This tissue fluid circulates in and around the brain.

Cerebrum This is the largest portion of the brain and consists of the left and right cerebral hemispheres. The cerebrum controls movement, sensation, learning, and memory.

Chemoreceptors Cells that respond to changes in their chemical environment by creating nerve impulses. Some chemoreceptors in the brain respond to carbon dioxide levels in the blood to help regulate breathing.

Chemotaxis The reaction of mobile cells to a chemical gradient; the cells may move either towards or away from the gradient depending on the nature of the chemical being used.

Chloride Shift Describes the exchange of negatively charged chloride ions for negatively charged bicarbonate ions across an erythrocyte's cell membrane.

Cholesterol A fatlike substance that occurs naturally in the body. Two types exist: high-density lipoprotein (HDL) and low-density lipoprotein (LDL).

Chondrocytes Cartilage cells.

Chordae Tendineae Tiny tendinous cords located at each of the heart valves. They attach to nearby muscles and prevent blood backflow through the valves.

Choroid Plexus This capillary network helps form the cerebrospinal fluid in the brain.

Chromatin A diffuse mixture of DNA and proteins that condenses into chromosomes prior to cell division.

Chromosomes Cellular structures composed of proteins and DNA that carry the body's hereditary information.

Cilia Hairlike projections from the surface of a cell. In the respiratory system, cilia help filter out foreign particles from the air before they reach the lungs.

Circadian Rhythm The body's 24-hour biological cycle that regulates certain activities, such as sleep, regardless of environmental conditions, including lightness and darkness.

Circle of Willis A vascular structure that supplies blood to the brain. It arises from the basilar, internal carotid, and other arteries.

Circulatory System The heart, blood vessels, and blood.

Circumcision The surgical removal of the foreskin. The term is also sometimes used with reference to females to describe a controversial and excruciating practice of genital mutilation that is common in certain societies around the world.

Citric Acid Cycle A chemical reaction that takes place in the mitochondria. The cycle produces some energy for the cell and produces products that can be used to produce large amounts of energy through oxidative phosphorylation.

Colic Arteries Divided into right, left, and middle colic arteries, all of which branch from either the inferior or superior mesenteric arteries, and feed the colon.

Collagen The albumin-like substance in connective tissue, cartilage, and bone.

Collecting Duct Where fluid is carried from the distal convoluted tubule (DCT) in the nephron of the kidneys on its way to the minor calyx.

Colostrum Nutritious fluids secreted by the breasts shortly before and after a woman gives birth; precedes the production of breast milk.

Common Carotid Arteries One of two major blood vessels that supply the head. The left carotid splits directly from the aortic arch between the bases of the two coronary arteries. The right carotid indirectly branches from the aorta via the brachiocephalic artery.

Complement The collective term for a variety of beta globulins. *See also* Globulins.

Complement Fixation The process by which complement factors bind to either antibodies or cell surfaces during the immune response.

Complementary Base Pair Nucleotide bases (adenine and thymine or guanine and cytosine) that pair up via hydrogen bonds in DNA.

Compression Forces Forces that squeeze items together; blows that press against the body.

Computed Tomography (CT) Scan A commonly used tool for determining the nature of a stroke.

Concentration Gradient The change in solute concentration from one location to another. Unless restricted, solutes will move from a site of higher solute
concentration to one of lower solute concentration, leading to an equilibrium between the two sites.

Concentric Contraction The type of contraction that occurs when a muscle contracts and grows shorter, such as the biceps muscle when bending the elbow.

Conchae Structures or parts that resemble a seashell in shape with three bony ridges or projections—the superior, middle, and inferior conchae—on the surface of the nasal cavity sides.

Contraceptive An agent that prevents ovulation, kills sperm, or blocks sperm from reaching the ovum for fertilization.

Convergence An impulse pathway where a neuron receives impulses from the nerve endings of thousands of other neurons but transmits its message to only a few other neurons.

Coronal Plane Divides the body into front and back portions.

Coronary Arteries Arising from the base of the aorta, these are the two major arteries that feed the heart muscle. The right coronary artery remains a single, large vessel, but the left coronary artery almost immediately splits into transverse and descending branches.

Coronary Circulation The circulatory system of the heart.

Corpus callosum A band of white matter connecting the cerebral hemispheres.

Corpus luteum Progesterone-secreting tissue that forms from a ruptured Graafian follicle in the mammalian ovary after the egg has been released.

Cortex The tissue layer that covers the brain.

Cortical Bone The hard, dense bone that forms the outer shell of all bones.

Corticotroph Cell in the anterior pituitary gland that secretes corticotropin (ACTH).

Cortisol A steroid hormone produced by the adrenal cortex that influences the body's stress response.

Cranial Nerves The brain's 12 pairs of nerves located in the peripheral nervous system.

Craniosacral Division Another name for the parasympathetic division of the autonomic nervous system. In this division, all the cell bodies of preganglionic neurons are located in the brain stem and sacral segments of the spinal cord.

Cranium The bones of the skull that house the brain.

Creatinine Waste produced by the breakdown of creatine phosphate in muscles.

Creatine Phosphate A molecule stored in the muscle that can quickly replenish ATP during a sudden burst of exercise.

CT Scan *See* Computed Tomography (CT) Scan.

Cuboid Bones Bones in the wrist that are shaped like cubes.

Cutaneous Senses The skin's sensory mechanisms whose receptors are located in the dermis.

Cytochromes A class of membrane-bound intracellular hemoprotein respiratory pigments. These enzymes function in electron transport as carriers of electrons.

Cytokines Signaling peptides secreted by immune cells and other types of cells in response to infection or other stimuli.

Cytoplasm Cellular material located between the nucleus and cell membrane.

Daughter Cells Cells arising from mitotic division that are identical to the parent cell.

Deglutition Another term used for the act of swallowing.

Dehydration Synthesis A form of chemical reaction that involves the removal of water to form a chemical bond. Also called a condensation reaction.

Dentin A tissue that is the majority of the mass of a tooth. It consists primarily of minerals (70 percent), with the remainder being water and organic material.

Depolarization When the electrical charges in a nerve cell reverse due to a stimulus. The rapid infusion of sodium ions causes a negative charge outside and a positive charge inside the cell membrane.

Detrusor Muscle Three layers of smooth muscle surrounding the mucosa of the bladder.

Diaphragm A muscle that aids in respiration. It separates the thoracic cavity from the abdominal cavity.

Diaphysis The central shaft of a bone.

Diastole The heart's resting period.

Diathroses Joints allowing free movement.

Diffusion The passive flow of molecules from one location to another.

Diploid Cells Any organism whose cells contain two copies of each chromosome. The majority of human cells, except sex cells and some liver cells, are diploid.

Dislocation Condition when a bone is moved out of a joint.

Distal Indicates direction away from the torso.

Distal Convoluted Tubule (DCT) Located between the loop of Henle and the collecting duct inside the nephron of the kidney.

Diuretic A substance (i.e., caffeine) that increases urine production.

Divergence An impulse pathway where a neuron receives impulses from a few other neurons and relays these impulses to thousands of other neurons.

DNA A nucleic acid that contains a cell's genetic or hereditary information.

Dopamine A neurotransmitter found in the motor system, limbic system, and the hypothalamus.

Dorsal Root The sensory root of a spinal nerve that attaches the nerve to the posterior part of the spinal cord.

Dorsal Root Ganglion An enlarged portion of the spinal nerve's dorsal root that contains the sensory neuron's cell bodies.

Down Syndrome Mental retardation associated with specific chromosomal abnormalities.

Dura Mater This fibrous connective tissue is the outermost layer of the brain's meninges.

Eccentric Contraction The type of contraction that occurs when a muscle contracts but the overall muscle grows longer rather than shorter; an eccentric contraction occurs in the biceps which contracts to control the arm as it extends, but the muscle grows longer rather than shorter.

Effector A muscle, gland, or other organ that responds after receiving an impulse.

Efferent Nerves Fibers leaving the central nervous system carrying messages to the muscles, joints, skin, or internal organs.

Efferent Neuron Nerve cells that carry impulses and messages away from the spinal cord and brain to the muscles and glands.

Efferent Vessels A vessel of the lymphatic or circulatory systems that carries fluid away from an organ or lymph node.

Eicosanoids Compounds derived from fatty acids that act like hormones to influence physiologic functions.

Elastin A protein of blood vessels that imparts elasticity.

Electrocardiogram (ECG or EKG) The product of an electrocardiograph, it is a printout depicting the heart's electrical activity. An ECG has five parts, each signified with the letter P, Q, R, S, or T, that reflect different phases in the heart activity.

Electrocardiograph (ECG or EKG) A device that records the heart's electrical activity as a jagged line on a sheet of paper, which is called an electrocardiogram.

Electrolyte A charged particle like calcium (Ca^{2+}) or magnesium (Mg^{2+}) that may have a number of functions in cells.

Electron Transport System A complex sequence found in the mitochondrial membrane that accepts electrons from electron donors and then passes them across the mitochondrial membrane creating an electrical and chemical gradient.

Embryogenesis The entire process of cell division and differentiation leading to the formation of an embryo.

End-diastolic Volume The amount of blood in a completely filled ventricle. In an adult, this is typically about 0.12 quarts (120 ml).

Endocardium The membrane lining the heart.

Endocrine System The body's organ system that controls hormone secretion.

Endothelium In blood vessels, it is also known as the tunica intima. The tunica intima forms the innermost layer of blood vessels.

Eosinophil A type of granulocyte that appears to be active in the moderation of allergic responses and the destruction of parasites.

Epiblast The outer layer of a blastocyst before differentiation into the ectoderm, mesoderm, or endoderm.

Epidemic A widespread outbreak of an infectious disease that affects a disproportionately large number of people within a given population.

Epinephrine *See* Adrenaline.

Epiphysis The portion of bone attached to another bone by a layer of cartilage.

Epithelial Cell A type of cell that lines organs and tissues of the body. It specializes in the exchange of materials with the external environment, such as the lumen of the gastrointestinal tract.

Epitope The specific area of an antigen to which the B cell receptor binds.

Equilibrium Balance mechanisms that are regulated by inner ear structures.

Erythroblast An early stage in red blood cell development.

Erythrocytes Red blood cells.

Erythropoietin A protein hormone produced by the kidneys that stimulates red blood cell production.

Estrogen Any of a family of hormones produced by the female ovaries that determine female sexual characteristics and influence reproductive development.

Excitatory Nerve/Fiber A nerve fiber that passes impulses on to other fibers.

Excitatory Synapse The passing of an impulse transmission to other synapses.

Exocrine Glands Glands that utilize ducts to release their secretions to the outside environment.

Extension A stretching out, as in straightening a limb.

External Urethral Sphincter Ring of voluntary muscle surrounding the end of the urethra, which regulates urine flow out of the body.

Extracellular Fluid The water found outside a cell that contains plasma and other tissue fluids.

Extrinsic Factors Another term used for vitamin B_{12} in the diet.

Facilitated Diffusion A passive process that utilizes a membrane-bound protein to move a compound across a membrane down its concentration gradient.

Fascia The connective tissue surrounding an entire muscle. The fascia becomes part of the tendon at either end of the muscle, connecting the muscle to the bone.

Fascicle A bundle of muscle fibers within the muscle surrounded by a tissue called the perimysium. Each muscle is made up of many fascicles.

Fast-twitch Muscle A type of muscle fiber that is able to contract very quickly. These fibers are predominantly found in muscles that must contract quickly and with great strength but do not need to contract over a long period of time.

Femoral Artery Arising from one of the two external iliac arteries, the femoral artery traverses the thigh to the popliteal artery.

Femoral Vein A large blood vessel in the thigh that collects blood from the popliteal vein and great saphenous vein and delivers it to the external iliac vein.

Fibrinogen A protein in plasma. It functions in blood clotting.

Fibular Vein *See* Peroneal Veins.

Flavoproteins The enzymes that contain flavin bound to a protein. Flavoproteins play a major role in biological oxidations.

Flexion The bending of a joint or of body parts having joints.

Follicle-stimulating Hormone (FSH) A hormone produced by the anterior pituitary gland that triggers sperm production in the testes and stimulates the development of follicles in the ovaries.

Fossa A hole or indentation.

Fossae The plural form of fossa.

Gametes Reproductive cells that, before fusing at fertilization, are haploid—they contain 23 instead of 46 chromosomes.

Gas Exchange In the respiratory system, gas exchange refers to the process of acquiring oxygen from the air and eliminating carbon dioxide from the blood.

Gastric Arteries Blood vessels of the digestive system. The left gastric artery stems from the celiac artery and supplies the stomach and lower part of the esophagus. The right gastric artery stems from the common hepatic artery and eventually connects with the left gastric artery.

Gastric Inhibitory Peptide (GIP) The gastrointestinal hormone whose main action is to block the secretion of gastric acid.

Gastric Veins Blood vessels of the digestive system. Blood from the stomach exits into the gastric veins, which then empty into a number of other veins that ultimately enter the portal vein (in the case of the left and right gastric veins) or the splenic vein (in the case of the short gastric vein).

Gastrin Hormone produced by the gastrointestinal system that regulates stomach acid secretion.

Gastroepiploic Arteries Blood vessels of the digestive system. The right gastroepiploic artery branches from the gastroduodenal artery. The left gastroepiploic artery branches from the splenic artery. Both provide blood to the stomach and duodenum.

Gene Expression In genetics, a term describing the results of activating of a gene.

Gene Transcription The process by which a strand of DNA is copied to form a complementary RNA strand.

Genetic Immunity A form of immunity to a pathogen that is inherited.

Genetic Imprinting Refers to differences in the way maternal or paternal genes are expressed in the offspring.

Genetic Sex Gender determination based on an XX or an XY chromosome configuration.

Glia Support cells in the brain.

Globulins Plasma proteins that function as transportation vehicles for a variety of molecules, in blood clotting and/or in the body's immune responses. They are
divided into three types alpha, beta, and gamma globulins.

Glomerular Capsule A cup-shaped sac that surrounds glomeruli of the nephrons.

Glomerular Filtrate The product of blood filtration in the nephrons of the kidneys.

Glomeruli Clusters of capillaries in the kidneys.

Glomerulus The singular form of glomeruli.

Glossopharyngeal Nerve The mixed nerve in the throat and salivary glands that contains sensory fibers for the throat and taste from the posterior one-third of the tongue.

Glucose A form of sugar that is a necessary component (along with oxygen) in cell respiration.

Glucose Tolerance Test A test measuring blood sugar levels that is often used to diagnose diabetes.

Glutamate A neurotransmitter associated with pain-related impulses.

Glycogen A storage form of carbohydrate. The liver converts fats, amino acids, and sugars to glycogen, which functions as a reserve energy supply for the body.

Glycogenolysis The breakdown of glycogen in the liver and in muscle tissue.

Glycolysis The process of breaking down glucose into two molecules of pyruvate. Glycolysis produces some energy for the cell and is the primary way of producing energy during anaerobic exercise.

Glycoprotein An organic compound composed of a joined protein and carbohydrate.

Goblet Cell An epithelial cell that secretes mucus.

Gonadotroph A cell in the anterior pituitary gland that secretes luteinizing hormone and follicle-stimulating hormone.

Gonadotropins Hormones (luteinizing hormone and follicle-stimulating hormone) released by the anterior pituitary gland that stimulate the ovaries and testes.

Graft Rejection The tendency of the immune system to reject transplanted tissue as foreign.

Granulocyte The most abundant type of white blood cell. *See also* Basophil, Eosinophil, and Neutrophil.

Gray Matter Nerve tissue located in the central nervous system containing cell bodies of neurons.

Growth Factors Proteins that act on cells to stimulate differentiation and proliferation.

Growth Hormone Hormone secreted by the anterior pituitary gland that promotes bone and muscle growth and metabolism.

Growth Hormone-Releasing Hormone (GHRH) A hormone that stimulates the anterior pituitary gland to secrete the growth hormone (GH).

Gyri Folds or ridges in the cerebral cortex.

H Zone The space between the two sets of actin filaments in the center of the sarcomere. The H zone grows smaller when the sarcomere contracts, and the actin filaments slide toward each other in the center of the sarcomere.

Haldane Effect A high concentration of oxygen, such as occurs in the alveolar capillaries of the lungs, that promotes the dissociation of carbon dioxide and hydrogen ions from hemoglobin.

HDL *See* High-density Lipoprotein (HDL).

Heart The muscular pump that powers the circulatory system.

Heart Attack Also known as a myocardial infarction, this condition happens when the supply of oxygen to a portion of the heart muscle is curtailed to such a degree that the tissue dies or sustains permanent damage.

Heart Failure A condition in which the heart can no longer carry out its pumping function adequately, resulting in slow blood circulation, poorly oxygenated cells, and veins that hold more blood.

Hematopoeisis The formation and maturation of blood cells.

Hematuria Blood in the urine.

Heme Group A ringlike chemical structure that is part of hemoglobin.

Hemes The deep-red organic pigment that contains iron and other atoms to which oxygen binds in blood hemoglobin. Hemes are found in most oxygen-carrying proteins.

Hemoglobin A large chemical compound in red blood cells that imparts their red color and also participates in transporting oxygen and carbon dioxide.

Hemolysis The rupture and destruction of red blood cells.

Hepatic Arteries The blood vessels supplying the liver and other organs. The common hepatic artery arises from the celiac trunk and supplies the right gastric, gastroduodenal, and proper hepatic arteries. The proper hepatic artery supplies the liver by way of the cystic artery.

Hepatic Portal System The name given to the portion of the circulatory system that connects the stomach and both intestines to the liver.

Hepatic Vein The blood vessel that collects blood from the liver and delivers it to the inferior vena cava.

High Blood Pressure A medical condition that arises when the pressure of the blood against the blood vessel walls exceeds normal limits. It results from a narrowing of the arterioles.

High-Density Lipoprotein (HDL) Often called the type of cholesterol curtails the accumulation of low-density lipoprotein in blood vessels.

Hilius The curved notch on the side of each kidney near the center where blood vessels enter and exit the kidney.

Histones Proteins associated with gene expression.

Homeostasis The regulation of the body's internal environment to maintain balance.

Homologous In genetics, chromosomes (one from the male parent, one from the female parent) carrying alleles for similar traits, such as eye color, that pair up during meiosis.

Hormone A chemical compound, often called a chemical messenger, that the brain and other organs use to communicate with the cells.

Huntington's Chorea A progressive and fatal disease affecting the nervous system.

Hydrolysis In chemistry, the breaking of a chemical bond by the addition of water.

Hydrophilic A water-loving compound, meaning that it is soluble in water. An example is glucose.

Hydrophobic A water-fearing compound, meaning that is it generally insoluble in water. Most lipids are hydrophobic, as are some amino acids.

Hypernatremia Too much sodium in the extracellular fluid.

Hypertension *See* High Blood Pressure.

Hypertrophy The process in which muscles grow larger in response to exercise.

Hyperventilation An increased and excessive depth and rate of breathing greater than demanded by the body's needs; can lead to abnormal loss of carbon dioxide from the blood, dizziness, tingling of the fingers and toes, and chest pain.

Hypoblast The inner layer of tissue in a developing embryo that will eventually become the digestive tract and respiratory tract.

Hypocalcemia A deficiency of calcium in the blood.

Hyponatremia Too little sodium in the extracellular fluid.

Hypophysis The pituitary gland.

Hypothalamic-Hypophyseal Portal System The circulation system through which neurohormones from the hypothalamus travel directly to the anterior pituitary gland without ever entering the general circulation.

Hypothalamic-Pituitary-Target Organ Axis A multiloop feedback system that coordinates the efforts of the hypothalamus, the pituitary gland, and the target gland.

Hypothalamus This part of the brain regulates body temperature and pituitary gland secretions. The hypothalamus is located superior to the pituitary gland and inferior to the thalamus.

Hypoxia A sudden decrease in the blood's oxygen content.

I Band The region between the Z band at the outside of the sarcomere and the end of the myosin chain that spans the center of the sarcomere.

Iliac Arteries These arise at the end of the abdominal artery. The abdominal artery bifurcates into two common iliac arteries, each of which soon divides again into internal and external iliac arteries.

Iliac Veins Blood from the femoral vein collects in the external iliac vein, which joins the internal iliac vein and carries blood from the pelvis to form the common iliac vein.

Immune System A body system that includes the thymus and bone marrow and lymphoid tissues. The immune system protects the body from foreign substances and pathogenic organisms in the form of specialized cellular responses.

Immunity The ability of an organism not to be affected by a given disease or pathogen.

Immunoglobulins (Ig) Plasma proteins that act as antibodies. The five main types are IgA, IgD, IgE, IgG, and IgM.

Inflammatory Mediators Soluble, diffusible molecules that act locally at the site of tissue damage and infection.

Inhibin Hormone secreted by the ovaries and testes that inhibits the release of follicle-stimulating hormone (FSH) by the pituitary.

Inhibitory Nerve A type of nerve fiber that obstructs impulse transmission to another fiber.

Inhibitory Synapse An impulse transmission obstruction due to a chemical inactivator located at the dendrite of the postsynaptic neuron.

In-series Blood Circulation Also known as portal circulation. It is blood flow that travels from one organ to another in series.

Insertion The end of the muscle that is usually farthest from the center of the body and usually the one that moves when the muscle contracts.

Insulin A hormone secreted by the pancreas. It allows the body cells to use energy, specifically glucose.

Insulin-like Growth Factors Substances produced in the liver and other tissues that act much like growth hormone, stimulating bone, cartilage, and muscle cell growth and differentiation.

Intercalated Disk A disk that separates two muscle fibers in the heart muscle. This disk can conduct the signal to contract from one muscle fiber to the next. With this connection, the entire heart muscle can contract in unison.

Intercostal Muscles Found under the ribs, these muscles play a role in respiration.

Interferons A family of drugs used to regulate the body's immune system. They may be used for such diseases as multiple sclerosis or cirrhosis of the liver.

Interlobar Arteries Blood vessels that branch from the renal artery to disperse blood throughout the kidney and to glomeruli.

Intermediate Pituitary A lobe of the pituitary of which only vestiges remain in humans.

Intermediolateral Cell Column Located on the thoracic level of the spinal cord, this is an extra cell column where all presynaptic sympathetic nerve cell bodies are located.

Internal Urethral Sphincter Ring of involuntary muscle that surrounds the urethra where it meets the bladder and that controls the flow of urine.

Intestinal Villi Tiny projections that line the inside wall of the small intestine and the uptake of nutrients by capillaries.

Intracapsular Ligaments Ligaments within the capsule at the joint.

Intracellular Fluid The water found within a cell.

Intrinsic Factors A protein released by the gastrointestinal tract that aids in the absorption of vitamin B_{12}.

In Vitro Occurring outside the body, often used to refer to laboratory procedures such as fertilization of ova within a laboratory dish.

In Vivo Occurring inside the body.

Involuntary Muscle *See* Smooth Muscle.

Ions Any element or compound that loses or gains electrons and in the process changes its net electric charge.

Islets of Langerhans Endocrine cells located in the pancreas in which the hormones insulin and glucagon are produced.

Isometric Contraction The type of contraction that occurs when a muscle contracts but the joint does not open or close, such as when pushing against a wall or pushing down on a table.

Joint The union between two bones.

Jugular Veins Blood vessels of the head and/or neck. The anterior jugular vein collects blood from veins of the lower face, traverses the front of the neck, and delivers the blood to the external jugular vein. The external jugular vein is a large vein that also receives blood from within the face and around the outside of the cranium, and empties into one of several veins, including the internal jugular. The internal jugular vein is the largest vein of the head and neck, and also drains blood from the brain and neck. It joins the subclavian vein to form the brachiocephalic vein.

Ketone Bodies Substances produced from fats when not enough glucose is present, which provide an alternate energy source for the brain and other tissues.

Kidneys The two bean-shaped organs that filter wastes, regulate electrolyte balance, and secrete hormones.

Kilocalorie The amount of energy required to raise 1,000 grams of water from 14.5° to 15.5° Celsius at standard atmospheric pressure.

Lacteals The portion of the lymphatic system that is associated with the gastrointestinal system, specifically the intestines.

Lactotroph A cell in the anterior pituitary gland that secretes prolactin.

Lateral Situated on a side.

LDL *See* Low-density Lipoprotein (LDL).

Leptin A protein hormone that influences metabolism and regulates body fat.

Ligament A tough band of connective tissue that connects bones to each other.

Lipoproteins Proteins that are connected chemically to lipids and used by the digestive system to transport hydrophobic fats and lipids in the hydrophilic bloodstream.

Loop of Henle The U-shaped section between the proximal convoluted tubule and the distal convoluted tubule in the nephron of the kidney.

Low-density Lipoprotein (LDL) Often called the "bad" cholesterol. This type of cholesterol can build up on blood-vessel walls and cause health problems.

Lumbar Veins Blood vessels of the digestive system. Lumbar veins collect blood from the abdominal walls and deliver it to other veins, including the inferior vena cava.

Lumen The internal diameter of a blood vessel. It represents the open space in the vessel through which the blood flows.

Luteinizing Hormone (LH) A hormone produced and secreted by the anterior pituitary gland that stimulates ovulation and menstruation in women and androgen synthesis by the testes in men.

Luteolysis The process by which the corpus luteum in the ovary degenerates when an egg is not fertilized.

Lymph Fluid in the vessels of the lymphatic system. It is the interstitial fluid that exits the capillaries and enters surrounding cells during the capillaries' exchange function.

Lymphatic System A series of vessels that shunts excess tissue fluid into the veins.

Lymph Node Filters that separate from lymph any invading organisms and other foreign materials.

Lymphocyte A type of leucocyte that detects antigens and serves in the body's immune response. The two main types are B cells and T cells.

Macrophage White blood cells that ingest and digest bacteria, other foreign organisms, platelets, and old or deformed red blood cells.

Magnetic Resonance Imaging (MRI) A diagnostic tool for viewing blood flow and locating sites of blood-flow blockage.

Major Calyx Openings in the center of the kidneys through which urine flows into the renal pelvis.

Malignant A condition that becomes progressively worse or more pronounced over time, and which may lead to death.

Medial Toward the midline of the body.

Medulla Located above the spinal cord, this part of the brain controls vital functions such as heart rate, respiration, and blood pressure.

Medullary Cords Within a lymph node, these are areas of dense lymphatic tissue.

Meiosis A process of cell division resulting in daughter cells containing half the number of chromosomes contained in the parent cell. In humans, this process is responsible for the generation of the sex cells, oocytes and sperm.

Meninges The membrane is composed of connective tissue that covers the brain and spinal cord and lines the dorsal cavity.

Mesenteric Arteries Blood vessels of the digestive system. The inferior and superior mesenteric arteries arise from the abdominal aorta and flow into numerous arteries of the large and small intestines, and the rectum.

Mesenteric Veins Blood vessels of the digestive system. The superior mesenteric vein drains the small intestine, and the inferior mesenteric collects blood from the colon and rectum. Both deliver their blood to the splenic vein.

Mesentery A tissue that suspends the digestive glands within the abdominal cavity. The mesentery connects to the outer layer of the gastrointestinal tract.

Metabolism The sum of all of the chemical reactions in a cell, tissue, organ, or organism. In nutritional terms, it frequently applies to the processing of the energy nutrients and generation of energy.

Microvilli Small outgrowths covering the intestinal villi. They increase the surface area of the villi, aiding in nutrient uptake by capillaries.

Micturition The process in which urine is released from the bladder; urination.

Midsagittal plane An imaginary line that passes through the skull and spinal cord, dividing the body into equal halves.

Mineralocorticoids A class of hormones produced by the adrenal cortex that regulate mineral metabolism.

Minor Calyx A cup-like receptacle attached to each renal pyramid in the kidney.

Mitochondria Located in the cell's cytoplasm, these are organelles where cell respiration takes place and energy is produced.

Mitosis A process of cell division resulting in daughter cells containing the same number of chromosomes as the parent cell.

Monocyte A type of white blood cell. They become macrophages, large cells that engage in phagocytosis.

Monozygotic Refers to twins arising from one ovum.

Morula A compacted group of embryonic cells at a level of development between the zygote and blastocyst stages.

Motilin A gastrointestinal hormone that stimulates intestinal muscle contractions to clean undigested materials from the small intestine.

Mucociliary Pertaining to mucus and to the cilia of the epithelial cells in the respiratory system.

Mucosa A mucous membrane that lines a body cavity.

Multiple Marker Test Testing to screen for various biomarkers of disease. *See also* Biomarker.

Muscle Fiber A muscle unit made up of many muscle cells that have fused together and received the signal to contract from a single nerve.

Muscle Spindle Related to the stretch reflex, this receptor responds to the muscle's passive stretch and contraction. The muscle spindles are parallel with the muscle fibers.

Myelin Sheath A substance composed of fatty material that covers most axons and dendrites in the central and peripheral nervous systems in order to electronically insulate neurons from one another.

Myofibril The contractile unit within a muscle fiber that is made up of a series of contractile units called sarcomeres. Each muscle fiber contains many myofibrils, all of which contract when the muscle fiber receives a signal from a nerve.

Myoglobin A molecule in the muscle that collects oxygen from the blood and delivers it to mitochondria in the muscle fiber.

Myosin One of the major contractile proteins making up a muscle fiber. Myosin proteins form chains that pull on actin filaments, causing the muscle fiber to contract.

Nephron The filtering unit of the kidney.

Nerve A system of neurons with blood vessels and other connective tissue.

Nerve Fiber The neuron including the axon and the surrounding cells. These fibers branch out at the neuron's ending, which is known as arborization.

Nerve Plexus A combination of neurons from various sections of the spinal cord that serve specific areas of the body.

Nerve Tracts A neuron group that performs a common function in the central nervous system. This grouping can be ascending (sensory) or descending (motor).

Neurohormone A chemical messenger released by the hypothalamus that signals the pituitary gland to release or inhibit release of its hormones.

Neurolemma Essential to the regeneration of damaged neurons in the peripheral nervous system, this is a sheath surrounding peripheral axons and dendrites and is formed by cytoplasm and the nuclei of Schwann cells.

Neuron A nerve cell that consists of a cell body, in addition to an axon and dendrites.

Neurosecretory Cells Specialized nerve cells that transmit chemical impulses, release hormones, and serve as a link between the endocrine and nervous systems.

Neurotransmitter Chemical substances that are emitted through nerve endings to help transmit messages. In the human body, there are about 80 different neurotransmitters.

Neutrophil The most common type of granulocyte. Neutrophils are a main bodily defense mechanism against infection, and are particularly suited to engulfing and destroying bacteria, although they can also combat other small invading organisms and materials.

Node of Ranvier The cell region located on or between the Schwann cells.

Noradrenalin A type of neurotransmitter that transports neurons throughout the various regions in the brain and spinal cord, in addition to increasing the reaction excitability in the CNS and the sympathetic neurons in the spinal cord.

Norepinephrine A hormone that causes blood pressure to rise in stressful situations.

Normoblast The cells of the bone marrow that are responsible for the formation of the red blood cells.

Nucleus The cell's largest organelle that contains chromosomes and hereditary material.

Occipital Lobes The most posterior part of the brain, containing the visual areas.

Oligodendrocytes A type of neuroglia that forms the neuron's myelin sheath.

Oocytes Ova that have not yet matured in the ovary; they arise from primordial oogonia that develop in the fetus.

Oogenesis The formation and development of an egg in the ovary.

Oogonia Cells that arise from primordial germ cells and differentiate into oocytes in the ovary.

Opposable Thumb In primates including humans, the ability to use the thumb to touch each finger.

Opsonization The modification of a bacterium so that it is more easily recognized by the immune system, resulting in an increase in phagocytosis by macrophages.

Organelles Primary components in a cell, including the nucleus, chromosomes, cytoplasm, and mitochondria.

Organic Molecules Molecules that contain carbon-carbon or carbon-hydrogen bonds.

Origin The end of the muscle closest to the body.

Osmoreceptors Neurons that sense fluid concentrations and send a message to the hypothalamus.

Osmosis A process that seeks to equalize the water-to-solute ratio on each side of a water-permeable membrane.

Osteology The study of bones, from the Greek word osteon, meaning "bone" and the suffix -ology, meaning "study of."

Ovarian Vein One of a pair of veins serving the female reproductive system.

Oxaloacetic acid An acid formed by oxidation of maleic acid, as in metabolism of fats and carbohydrates in the Krebs cycle.

Oxidation-reduction Reaction A reaction in which there is transfer of electrons from an electron donor (the reducing agent) to an electron acceptor (the oxidizing agent). Also called the redox reaction. In the electron transport system, this reaction results in molecules alternately losing and gaining an electron.

Oxidative Phosphorylation The process of combining electrons with oxygen to create water. This process also produces energy for the cell when enough oxygen is present.

Oxidization Add oxygen to or combine with oxygen, usually in chemical processes.

Oxygen Dissociation Curve A graph that shows the percent saturation of hemoglobin at various partial pressures of oxygen. The curve shifts to the right (the Bohr effect) when less than a normal amount of oxygen is taken up by the blood and shifts to the left (the Haldane effect) when more than a normal amount is taken up.

Pacemaker *See* SA Node.

Palmar Indicates the palms of the hands.

Pancreatic Polypeptide Hormone secreted by the F cells of the endocrine pancreas that inhibits gallbladder contraction and halts enzyme secretion by exocrine cells in the pancreas.

Pandemic An epidemic that occurs over a large geographic area, sometimes throughout the world.

Papillary Duct A tube that drains urine from collecting ducts in the nephron and empties it into the minor calyx.

Paracrine The action of a hormone on neighboring cells.

Parasympathetic Nervous System Also known as the vagal system. It is one of two major divisions of the autonomic nervous system. It functions to inhibit the pacemaker and lower the heart rate. *See also* Sympathetic Nervous System.

Parathyroid Hormone (PTH) A hormone secreted by the parathyroid gland that helps maintain calcium and phosphorous levels in the body. PTH controls the release of calcium from bone, the absorption of calcium in the intestine, and the excretion of calcium in the urine. Also called parathormone.

Partial Pressure Within the circulatory system, it is a term used to describe the relative oxygen concentration in tissues. For example, hemoglobin has a differential ability to bind oxygen: It picks up oxygen when the partial pressure in surrounding tissues is high, as it is in the lungs, and drops off oxygen when the partial pressure in the surrounding tissues is low, as it is in the tissues.

Pathogens Disease-producing agents such as virus, bacterium, or other microorganisms.

Peptides A chemical that helps to join amino acids in a protein molecule.

Pericardium The two-layered membranous sac around the heart.

Perimysium Connective tissue that surrounds the bundle of muscle fibers making up a fascicle.

Peripheral Nervous System Division of the nervous system that consists of the spinal and cranial nerves.

Peristaltic Action A rhythmic contraction of the muscles of the gastrointestinal tract, most notably in the small intestine, that is responsible for moving nutrients and undigested material through the lumen towards the anus.

Peroneal Veins Also known as fibular veins. They drain the lower leg and ankle, and deliver the blood to the posterior tibial vein.

pH The acidity of a solution. It is formally the measure of the hydrogen ion concentration of a solution.

Phagocytic Cell A type of cell that engulfs external particles, food, or organisms into its cytoplasm; the enclosed material may then be destroyed by digestive enzymes.

Phagocytosis The process of engulfing and destroying bacteria and other antigens.

Pharynx The rear area of the oral cavity. This area connects the respiratory and digestive systems of the body.

Phosphate A chemical related to energy usage and transmission of genetic information in the cell.

Phospholipids A class of organic molecules that resemble triglycerides but have one fatty acid chain replaced by a phosphate group.

Pia Mater The meninges' innermost layer, made of thin connective tissue located on the surface of the brain and spinal cord.

Pituitary Gland An endocrine gland at the base of the brain that sends out growth hormones.

Plantar Indicates the soles of the feet.

Plasma The liquid portion of blood in which red and white blood cells, platelets, and other blood contents float.

Plasminogen A beta globulin that participates in blood clotting.

Plasticity The reorganization of the nervous system following an injury or a tissue-damaging disease.

Platelets Also known as thrombocytes. They are round or oblong disks in the blood that participate in blood clotting.

Pleura A membrane that envelops the lung and attaches the lung to the thorax. There are two pleurae, right and left, that are entirely distinct from each other, and each pleura is made of two layers. The parietal pleura lines the chest cage walls and covers the upper surface of the diaphragm, and the visceral pleura tightly covers the exterior of the lungs. The two layers are actually one continuous sheet of tissue that lines the chest wall and doubles back to cover the lungs. The pleura is moistened with a thin, serous secretion that helps the lungs to expand and contract in the chest.

Polarization A chemically charged state when the neuron's membrane has a positive charge outside and a negative charge inside.

Polyploid Cells In humans, each cell has two copies of each chromosome, one maternal and one paternal. Polyploid indicates more than two chromosomes in a cell.

Polyspermy The entrance of several sperm into an ovum.

Polyunsaturated Fatty Acids Components of dietary fats that contain at least two double bonds.

Pons The parts of the brain that are anterior and superior to the medulla. The pons regulate respiration.

Popliteal Artery A blood vessel that arises from the femoral artery and traverses the knee before dividing into the posterior and anterior tibial arteries.

Popliteal Vein A blood vessel that collects blood from the anterior and posterior tibial veins, and empties into the femoral vein.

Porphyrin A complex, nitrogen-containing compound that makes up the various pigments found in living tissues. Iron-containing porphyrins are called hemes.

Portal Circulation *See* In-series Blood Circulation.

Portal Vein A blood vessel that arises from the splenic vein and superior mesenteric vein. It empties into the liver.

Positron Emission Tomography (PET) Scan A type of brain imaging technique that shows the brain in action. In order to obtain this image, a radioactive substance (such as glucose) is injected into the brain and then followed as it moves throughout the brain.

Posterior Indicates the back of a person or mammal.

Posterior Pituitary Lobe of the pituitary gland that is an extension of the nervous system.

Postganglionic Neuron A neuron located in the autonomic nervous system that extends from a ganglion to the visceral effector.

Postsynaptic Any impulse event following transmission at the synapse.

Preganglionic Neuron A neuron located in the autonomic nervous system that extends from the central nervous system to a ganglion and then synapses with a postganglionic neuron.

Pregnenolone A steroid hormone precursor produced from cholesterol.

Preprohormone/Prohormone An inactive sequence of amino acids from which an active hormone is released.

Progesterone Steroid hormone produced in the adrenal gland, placenta, and corpus luteum that influences sexual development and reproduction.

Progestin Female hormone produced by the ovaries that influences sexual development and pregnancy.

Proglucagon Precursor molecule from which the hormone glucagon is produced.

Proinsulin The inactive precursor molecule from which insulin is formed.

Projection A sensory occurrence when the sensation is felt in the receptor area.

Prolactin A protein hormone secreted by the anterior pituitary that stimulates mammary gland development and milk production.

Prostaglandin Fatty acid derivatives that act much like hormones to influence a number of physiological processes throughout the body.

Prostate The gland surrounding the top of the urethra in men that contributes nutrients to the seminal fluid.

Protease A class of enzyme that is involved in the breakdown of proteins into amino acids.

Protein Complex chemical compounds that are essential to life.

Prothrombin A beta globulin that participates in blood clotting.

Protozoa Single-celled, eukaryotic organisms, including many parasites.

Proximal Indicates direction closer to the torso.

Proximal Convoluted Tubule (PCT) Tiny tubes in the nephrons of the kidneys through which glomerular filtrate passes and substances necessary to the body (i.e., water, sodium, and calcium) are reabsorbed into the bloodstream.

Pulmonary Artery The blood vessel that originates at the right ventricle, then splits into two branches. The left and right pulmonary arteries lead to the left and right lung, respectively.

Pulmonary Circulation The transit of blood from the heart to the lungs and back to the heart. Blood picks up oxygen and drops off carbon dioxide in this circulatory route.

Pulmonary Semilunar Valve The three-cusped heart valve located between the right ventricle and pulmonary artery.

Pulmonary Veins Four blood vessels that flow from the lungs to the left atrium.

Purkinje Fibers A mesh of modified muscle fibers located in the base of the ventricle wall. The fibers receive the electrical impulse from the bundle of His and deliver it to the ventricle, which then contracts.

Pyruvate The end product of glycolysis.

Radial Artery A blood vessel in each lower arm that receives blood from the brachial artery and delivers it to numerous arteries of the forearm, wrist, and hand.

Radial Vein A blood vessel in each arm that collects blood from veins in the hand. It eventually merges with the ulnar vein into the brachial vein.

Receptors Proteins on the surface of cells or within cells that bind to particular hormones.

Rectal Vein Blood vessels in the digestive system that drain parts of the rectum. The inferior rectal vein joins the internal pudendal vein, which flows into the internal iliac vein, while the middle rectal vein connects directly to the internal iliac vein. The superior rectal vein flows directly into the inferior mesenteric vein.

Red Blood Cells Also known as erythrocytes. These are the cells in the blood that are responsible for gathering and delivering oxygen and nutrients to the body tissues, and for disposing of the tissue's waste products.

Reflex An automatic or involuntary response to a stimulus.

Renal Artery A pair of blood vessels that arise from the abdominal aorta. Each feeds a kidney and adrenal gland, and the ureter.

Renal Fascia The outermost layer of the kidney, composed of connective tissue that holds the kidney to the abdominal wall.

Renal Pelvis A funnel-shaped cavity that collects urine and sends it into the ureter.

Renal Pyramids Cone-shaped receptacles inside the medulla of the kidney.

Renal Veins A pair of blood vessels that drain the two kidneys. They empty into the inferior vena cava.

Renin An enzyme secreted by the kidneys that leads to the production of the hormone aldosterone.

Repolarization A chemically charged state following a neuron's depolarization, when the membrane has a positive charge outside and a negative charge inside due to the outflow of potassium ions.

Respiration In the respiratory system, the movement of respiratory gases, such as oxygen and carbon dioxide, into and out of the lungs.

Reticulocyte Immature red blood cells; these are usually found in the bone marrow.

Rh Factor An antigen that is found on the surface of blood cells; it is an independent factor of the ABO group.

Rotation Involves turning a body part on an axis.

Saggital Plane An imaginary vertical line that divides the body into right and left segments.

SA Node Also known as the sinoatrial node, or pacemaker. This is a group of small and weakly contractile modified muscle cells that spontaneously deliver the electrical pulses that trigger the heart's contraction.

Sarcolemma The cell membrane of a muscle fiber.

Sarcomere An individual contractile unit within the myofibril that contains actin filaments attached to either end. Myosin chains pull the actin filaments closer together, making the sarcomere grow shorter.

Sarcoplasmic Reticulum A network of tubules that runs throughout the muscle fiber. The sarcoplasmic reticulum stores calcium when the fiber is not contracted and releases calcium when the fiber receives a signal to contract.

Schwann Cells Located in the peripheral nervous system, these cells form the myelin sheath and neurolemma of the peripheral axons and dendrites.

Semilunar Valves Valves, shaped like half-moons, that ensure blood movement in only one direction. They are found in the heart and in large blood vessels.

Seminiferous Tubules Tubes in the testes in which sperm are produced.

Semipermeable (or Selectively Permeable) Membrane A membrane that allows certain molecules to pass through while restricting others.

Sensory Nerves A type of afferent nerve coming in at the back of the spinal cord; also called posterior nerves.

Sensory Neurons Also known as afferent neurons, they carry impulses and messages to the spinal cord and brain.

Septum A partition, dividing wall, or membrane that separates bodily spaces or masses of tissue. In the respiratory system, septum most often refers to the cartilage separating the two nostrils.

Serosa The outer layer of the bladder wall.

Serotonin A neurotransmitter present throughout the central nervous system.

Sertoli Cells Cells in the testes in which sperm is produced.

Sesamoid Bone Short bones embedded within a tendon or joint capsule.

Sex-linked Inherited Characteristics Traits, such as color-blindness, that are linked to genes on the sex chromosomes, especially the X chromosome.

Sickle Cell Anemia A serious autosomal recessive disease characterized by abnormal red blood cells.

Sigmoidal Artery A blood vessel that arises from the inferior mesenteric artery and supplies blood to the lower abdominal region.

Skeletal Muscle Muscles that are attached to the skeleton and allow the body to move. This is also called voluntary muscle because these are the muscles that move voluntarily.

Slow-twitch Muscles A type of muscle fiber that is able to contract very quickly. These are predominantly found in muscles that must contract repeatedly but without much strength.

Smooth Muscle Also known as an involuntary muscle. It is a type of muscle that is controlled by the autonomic nervous system, rather than by willful command, as is the striated muscle.

Sodium/Potassium Pump A form of active transport that regulates the amount of sodium and potassium in and around the cells.

Somatic Neuron A type of sensory neuron located in the skeletal muscle and joints.

Somatostatin A hormone produced by the endocrine pancreas and hypothalamus that regulates insulin and glucagon release, and inhibits growth hormone release from the pituitary gland.

Somatotroph A cell in the anterior pituitary gland that secretes growth hormone.

Spermatogonia Primordial sperm cells that develop in the male fetus.

Sphincter A skeletal muscle that forms a circular band and that usually controls the size of an opening, such as the mouth or the entrance to the stomach. The muscle contracts to close the opening or relaxes to open it.

Spinal Nerves The spine's 31 pairs of nerves located in the peripheral nervous system.

Spinal Reflex An automatic or involuntary reflex related to the spinal cord and in which the brain is not directly involved.

Splenic Artery Blood vessel that arises from the celiac trunk and branches into numerous arteries that feed the stomach and peritoneum, pancreas, and spleen.

Splenic Vein A large blood vessel that collects blood from the spleen. It joins the superior mesenteric vein to create the portal vein.

Stem Cells Undifferentiated cells. They have the genetic potential to mature into specific cell types. Some stems are only able to become one type of cell, while others have the ability to become any number of different cells.

Stimulus Any sort of change in a living organism that causes a response or affects a sensory receptor.

Stretch Reflex A reflex from the spinal cord in which a muscle will respond to a stretch by contracting.

Striated Muscle Also known as voluntary muscle. A person can consciously control the action of striated muscle.

Subclavian Arteries Blood vessels that supply the arms, much of the upper body, and the spinal cord. The right subclavian artery branches from the brachiocephalic artery, while the left divides off of the aortic arch. Numerous arteries arise from each.

Subclavian Veins Primary blood vessels draining the arms. They collect blood from the axillary vein and later merge with the internal jugular vein to produce the brachiocephalic vein.

Substance P Neuropeptide found in the gut and brain that stimulates smooth muscle contraction and epithelial cell growth and that plays a role in both the pain and pleasure responses.

Sulci Grooves between the gyri of the cerebellum.

Superior Direction given to a body part that indicates toward the head.

Surfactant A substance that acts on the surface of objects. In the respiratory system, surfactants are secreted by pneumocyte cells into the alveoli and respiratory air passages, helping to make pulmonary tissue elastic in nature.

Sympathetic Nervous System One of two major divisions of the autonomic nervous system. It functions to stimulate the pacemaker and boost the heart rate. *See also* Parasympathetic Nervous System.

Symphysis A disk of cartilage where two bones meet fiber that attaches a muscle to a bone.

Synapse The junction between two neurons where the axon passes on information to the dendrite. This area is often called a relay because it is here where the information is relayed to the next neuron.

Synaptic Gap or Cleft The actual area (which is approximately 10–50 nanometers in width) between the axon and dendrite where the neurons communicate with each other.

Synarthroses Nonmoveable joints.

Synergist A muscle that works in conjunction with an antagonistic pair to control the movement of a joint. The synergist usually runs beside a joint or diagonally across a joint.

Synovial Fluid The clear fluid that is normally present in joint cavities.

Systemic Circulation The transit of blood from the heart to the body (except the lungs) and back to the heart. *See also* Coronary Circulation and Pulmonary Circulation.

T Cells Also known as T lymphocytes. They are one of two main types of lymphocyte, and participate in the body's immune response.

T Tubule Tubules that run through muscle fibers carrying the signal to contract. The signal passes from the T tubule to the sarcoplasmic reticulum, which releases calcium and causes the contraction to take place.

Target Cells Cells that are responsive to a particular hormone.

Tendon A band of connective tissue that connects the muscle to the bone.

Terminal Arterioles Arterioles that feed capillaries.

Testosterone A hormone that produces male characteristics including large muscles.

Tetanus Contraction A sustained contraction as a result of many independent signals from a nerve.

Thalamus The portion of the brain located superior to the hypothalamus that controls the elements of subconscious sensation.

Threshold Level This value in a nerve fiber depends on the composition of the cellular fluid and the number of impulses recently received and conducted. When this level is reached in the nerve fiber's axon, a reaction results.

Thrombocytes *See* Platelets

Thromboplastin A substance released by damaged tissue and platelets. With calcium, it promotes the formation of blood clots.

Thyroid-stimulating Hormone (TSH) Hormone produced by the pituitary gland that stimulates the thyroid gland to secrete its hormones, thyroxine (T4) and triiodothyronine (T3). Also called thyrotropin.

Thyrotroph Cell in the anterior pituitary gland that secretes thyroid-stimulating hormone.

Thyrotropin-releasing Hormone (TRH) Hypothalamic neurohormone that triggers the release of thyroid-stimulating hormone (TSH) and prolactin (PRL) from the pituitary gland.

Thyroxine (T4) Thyroid hormone that influences metabolism and growth.

Tibial Arteries Blood vessels of the lower leg. The posterior and anterior tibial arteries arise from the popliteal artery and supply blood to arteries feeding the lower leg, ankle, and foot.

Tibial Veins Blood vessels of the lower leg. The anterior and posterior tibial veins drain the leg, then join together to form the popliteal vein.

Tonsils The name given to the lymphatic tissue found at the back of the oral cavity.

Toxoid The toxin produced by a bacterium that has been detoxified, but still retains its antigen characteristics. Toxoids are useful in the generation of immunizations.

Trabeculae Beams that act as strengthening girders of cancellous bone.

Trabecular Bone The porous, spongy bone that lines the bone marrow cavity and is surrounded by cortical bone.

Transverse Plane An imaginary line passing at right angles to both the front and midsection; a cross section.

Trigone A triangular-shaped region located in the bladder floor.

Triiodothyronine (T3) The more potent of the two thyroid hormones.

Tropomyosin A protein that forms long filaments wrapping around actin within the muscle fiber.

Troponin A protein that is associated with actin and tropomyosin within the muscle fiber.

Tunica Adventitia Fibrous connective tissue forming the outer of the three layers comprising arteries, arterioles, veins, and venules. *See also* Tunica Intima and Tunica Media.

Tunica Intima Also known as endothelium. It forms the innermost of the three layers comprising arteries, arterioles, veins, and venules. Capillaries are composed of only a single layer of endothelial cells. *See also* Tunica Adventitia and Tunica Media.

Tunica Media Muscular and elastic tissue forming the middle of the three layers comprising arteries, arterioles, veins, and venules. *See also* Tunica Adventitia and Tunica Intima.

Type A Blood Blood containing a certain antigen called "A." Due to potential antigen reactions, a person with type A blood can receive blood donations of type A and type O, but not type B or type AB.

Type AB Blood Blood containing anti-lymphocytes. *See also* T Cells.

Type B Blood Blood containing a certain antigen called "B." Due to potential antigen reactions, a person with type B blood can receive blood donations of type B and type O, but not type A or type AB.

Type O Blood Blood containing neither of the antigens called "A" and "B." Due to potential antigen reactions, a person with type O blood can receive blood donations of type O, but not type A, type B, or type AB.

Tyrosine An amino acid component of protein.

Ultrasound Scan An imaging method using high-frequency sound waves to form images inside the body. Also called ultrasonography.

Urea Waste produced by the breakdown of proteins.

Ureter A long tube that delivers urine from the kidney to the bladder.

Ureteral Orifices Two holes where the ureters pierce the bladder.

Urethra A muscular tube that connects the bladder with the exterior of the body.

Uric Acid Waste produced by the breakdown of nucleic acids (DNA and RNA).

Urochrome Pigment produced by the breakdown of bile that gives urine its yellow or amber color.

Vagus Nerve The 10th of 12 cranial nerves, which originates somewhere in the medulla oblongata in the brainstem and extends down to the abdomen.

Vasoconstrictor Nerves Nerves that signal the veins to constrict.

Vasodilation The relaxation of the muscles surrounding the vascular tissue; this increases the diameter of the vessel and reduces pressure.

Vasopressin A hormone produced by the pituitary gland that increases the permeability of the kidney ducts to return more fluid to the bloodstream. Also called antidiuretic hormone (ADH).

Vena Cava One of two large veins, the superior and inferior venae cavae, bringing blood from the body back to the heart.

Ventral Root The motor root of a spinal nerve that attaches the nerve to the anterior part of the spinal cord.

Ventricle In the human heart, it is one of the heart's two lower chambers.

Vertebral arteries A pair of blood vessels on each side of the neck that arise from the subclavian arteries. They unite at the basilar artery.

Vestibule The opening or entrance to a passage, such as the vestibule of the vagina.

Vestigial A term for nonfunctional remnants of organs.

Virus A nonliving infectious agent that is characterized as having a protein covering and either DNA or RNA as its genetic material; some viruses may also have a lipid covering. Viruses are completely dependent on cells for reproduction.

Visceral Neuron A type of sensory neuron located in the body's internal organs.

Visceral Organs The body's internal organs, such as the heart and lungs, that have nerve fibers and nerve endings that conduct messages to the brain and spinal cord.

White Blood Cells Also known as leukocytes. These are the cells in the blood that function in the body's defense mechanism to detect, attack, and eliminate foreign organisms and materials.

White Matter The nerve tissue located within the central nervous system that contains myelinated axons and interneurons.

Z Band A dense area that separates the sarcomeres. The actin filaments are embedded in the Z band, extending inward into each sarcomere.

Zona Fasciculate The middle layer of the adrenal cortex, in which the glucocorticoids (cortisol) are produced.

Zona Glomerulosa The outermost layer of the adrenal cortex, in which the mineralocorticoids (aldosterone) are produced.

Zona Pellucid The outer covering of an ovum.

Zona Reticularis The innermost layer of the adrenal cortex, in which the gonadocorticoids (sex hormones) are produced.

Zygote A diploid cell resulting from fertilization of an egg by a sperm cell.

Select Bibliography

Aaronson, Philip I., and Jeremy P. T. Ward, with Charles M. Wiener, Steven P. Schulman, and Jaswinder S. Gill. *The Cardiovascular System at a Glance*. Oxford: Blackwell Science Limited, 1999.

Abrahams, Peter, ed. *How the Body Works*, London: Amber Books, 2009.

"Acne." National Institute of Arthritis and Musculoskeletal and Skin Diseases. http://www.niams.nih.gov/Health_Info/Acne/default.asp (accessed June 20, 2010).

Adams, Amy. *The Muscular System*. Westport, CT: Greenwood Publishing, 2004.

"Alcohol-Induced Liver Disease." American Liver Foundation, http://www.liverfoundation.org/abouttheliver/info/alcohol/ (accessed June 20, 2010).

American Academy of Allergy, Asthma and Immunology. http://www.aaaai.org (accessed June 20, 2010).

American Academy of Family Physicians. http://www.familydoctor.org (accessed June 20, 2010).

American Academy of Otolaryngology. http://www.entnet.org (accessed June 20, 2010).

Asimov, Isaac. *The Human Body: Its Structure and Operation*. Rev. ed. New York: Mentor, 1992.

Bainbridge, David. *Making Babies: The Science of Pregnancy*. Cambridge, MA: Harvard University Press, 2001.

"Bariatric Surgery for Severe Obesity." National Institute of Diabetes and Digestive and Kidney Diseases, Weight-Control Information Network. http://win.niddk.nih.gov/publications/gastric.htm (accessed June 20, 2010).

Bastian, Glenn F. *An Illustrated Review of the Urinary System.* New York: HarperCollins College Publishers, 1994.

Berne, Robert M., and Matthew N. Levy. *Cardiovascular Physiology.* 6th ed. St. Louis, MO: C. V. Mosby-Year Book, 1992.

Charlton, C. A. C. *The Urological System.* Harmondsworth, UK: Penguin Books, 1973.

Cornett, Frederick D., and Pauline Gratz. *Modern Human Physiology.* New York: Holt, Rinehart, and Winston, 1982.

"Did You Know . . . Facts about the Human Body." Health News. http://www.healthnews.com (accessed June 20, 2010).

"Drinking Water." Centers for Disease Control and Prevention. http://www.cdc.gov/healthywater/drinking/travel/index.html (accessed June 20, 2010).

"Flu." Centers for Disease Control and Prevention. http://www.flu.gov (accessed June 20, 2010).

"Fun Science Facts." High Tech Science. http://www.hightechscience.org/funfacts.htm (accessed June 20, 2010).

Gilbert, S. F., M. S. Tyler, and R. N. Kozlowski. *Developmental Biology,* 6th ed. Sunderland, MA: Sinauer Associates, 2000.

"Global Water, Sanitation, and Hygiene (WASH)." Centers for Disease Control and Prevention. http://www.cdc.gov/healthywater/global/index.html (accessed June 20, 2010).

Greenspan, Francis S., and David G. Gardner. *Basic and Clinical Endocrinology.* 6th ed. New York: Lange Medical Books/McGraw-Hill, 2001.

"The Heart: An Online Exploration." http://sln.fi.edu/biosci/heart.html (accessed June 20, 2010).

Hess, Dean, and Robert M. Kacmarek. *Essentials of Mechanical Ventilation.* 2nd ed. New York: McGraw-Hill, Health Professions Division, 2002.

Hlastala, Michael P., and Albert J. Berger. *Physiology of Respiration.* 2nd ed. New York: Oxford University Press, 2001.

Hollen, Kathryn. *The Reproductive System.* Westport, CT: Greenwood Publishing, 2004.

Holmes, Oliver. *Human Neurophysiology: A Student Text.* 2nd ed. London: Chapman & Hall Medical, 1993.

"How Does Smoking Affect the Heart and Blood Vessels?" National Heart and Lung Institute. http://www.nhlbi.nih.gov/health/dci/Diseases/smo/smo_how.html (accessed June 20, 2010).

"The Human Body." Teachnology. http://www.teach-nology.com/themes/science/humanb/ (accessed June 20, 2010).

"Interesting Facts about the Human Body." Random Facts. http://facts.randomhistory.com/2009/03/02_human-body.html (accessed June 20, 2010).

Kelly, Evelyn. *The Skeletal System.* Westport, CT: Greenwood Publishing, 2004.

Knight, Bernard. *Discovering the Human Body.* New York: Lippincott & Crowell, 1980.

"LASIK." Food and Drug Administration. http://www.fda.gov/MedicalDevices/ProductsandMedicalProcedures/SurgeryandLifeSupport/LASIK/default.htm (accessed June 20, 2010).

Lyman, Dale. *Anatomy DeMystified.* New York: McGraw-Hill, 2004.

"Massage Therapy: An Introduction." National Center for Complementary and Alternative Medicine. http://nccam.nih.gov/health/massage/ (accessed June 20, 2010).

McDowell, Julie. *The Nervous System and Sensory Organs.* Westport, CT: Greenwood Publishing, 2004.

McDowell, Julie, and Michael Windelspecht. *The Lymphatic System.* Westport, CT: Greenwood Publishing, 2004.

"Medical References." University of Maryland Medical Center. http://www.umm.edu/medref/ (accessed June 20, 2010).

Mertz, Leslie. *The Circulatory System.* Westport, CT: Greenwood Publishing, 2004.

"National Cholesterol Education Program." National Heart Lung and Blood Institute. http://www.nhlbi.nih.gov/chd/ (accessed June 20, 2010).

"National Diabetes Statistics, 2007." National Institute of Diabetes and Digestive and Kidney Diseases. http://diabetes.niddk.nih.gov/dm/pubs/statistics/index.htm#what (accessed June 20, 2010).

National Institute of Allergy and Infectious Diseases. http://www.niaid.nih.gov (accessed June 20, 2010).

Northwestern University Medical School, Department of Neurology. http://www.neurology.northwestern.edu/ (accessed June 20, 2010).

Petechuk, David. *The Respiratory System*. Westport, CT: Greenwood Publishing, 2004.

Phillips, Chandler A., and Jarold S. Petrofsky. *Mechanics of Skeletal and Cardiac Muscle*. Springfield, IL: Thomas, 1983.

Sanders, Tina, and Valerie C. Scanlon. *Essentials of Anatomy and Physiology*. 3rd ed. Philadelphia: F. A. Davis Company, 1999.

Sherwood, Lauralee. *Human Physiology: From Cells to Systems*. 4th ed. Pacific Grove, CA: Brooks/Cole, 2001.

Soloman, Eldra P., Linda R. Berg, Diana W. Martin, et al. *Biology*. 4th ed. Orlando, FL: Harcourt Brace & Company, 1997.

"Spinal Cord Research." Christopher and Dana Reeve Foundation. http://www.christopherreeve.org/site/c.ddJFKRNoFiG/b.4343879/k.D323/Research.htm (accessed June 20, 2010).

"Sports Injuries." National Institute of Arthritis and Musculoskeletal and Skin Diseases. http://www.niams.nih.gov/Health_Info/Sports_Injuries/default.asp (accessed June 20, 2010).

Steele, D. Gentry, and Claude A. Bramblett. *The Anatomy and Biology of the Human Skeleton*. College Station: Texas A&M University Press, 1988.

Takahashi, Takeo. Atlas of the Human Body. New York: HarperCollins Publishers, 1989.

Watson, Stephanie. *The Endocrine System*. Westport, CT: Greenwood Publishing, 2004.

Watson, Stephanie. *The Urinary System*. Westport, CT: Greenwood Publishing, 2004.

"What Is Coronary Disease?" National Heart and Lung Institute. http://www.nhlbi.nih.gov/health/dci/Diseases/Cad/CAD_WhatIs.html (accessed June 20, 2010).

Windelspecht, Michael. *The Digestive System*. Westport, CT: Greenwood Publishing, 2004.

Index

A bands, 339–40
Abdomen, 20–21, 335
Abdominal aorta, 47
Abdominal cavity, 17
Abduction, 557–58
ABO blood type group,
    252–54
Abortion, 495–96
Acetabulum, 584–85
Acetylcholine, 195, 342, 394–95,
    433–34
Acetyl coenzyme A (acetyl CoA),
    354–55, 361, 374, 524
Acetylsalicylic acid, 118
Acidosis, 616
Acinar cells, 148
ACL (anterior cruciate
    ligament), 595
Acne, 219–20
Acquired immunity, 267–71
ACTH. *See* Adrenocorticotropic
    hormone
Actin, 338–40, 344–47,
    349–50, 369
Action potential, 395, 399
Active immunity, 267–69
Adaptation, 436

Adaptive responses, 258–59
Adduction, 558
Adenine, 9–11, 457
Adenoid, 257
Adenosine diphosphate (ADP),
    344, 352, 359, 386, 526
Adenosine monophosphate
    (AMP), 526
Adenosine triphosphate (ATP)
  cell respiration and, 6, 386,
    521–26
  muscle contraction and,
    344–48, 351–61, 363–65,
    367–68, 374
  neurotransmitters, 394, 521
  sodium/potassium pump
    and, 607
ADH. *See* Antidiuretic hormone
Adipocytes, 559
Adipose capsule, 602
Adipose tissue, 7, 14
ADP. *See* Adenosine diphosphate
Adrenal cortex, 188–90
Adrenal glands, 151–52,
    185–90, 432
Adrenaline, 393
Adrenal medulla, 185–88